SCHA

THEORY

OF

PROBABI

•

BY

SEYMOUR LIPSCHUTZ, Ph.D.
Associate Professor of Mathematics
Temple University

•

SCHAUM'S OUTLINE SERIES
McGRAW-HILL BOOK COMPANY
New York, St. Louis, San Francisco, Sydney

ISBN 07-037982-3

8 9 10 11 12 13 14 15 SH SH 7 5 4

Preface

Probability theory had its beginnings in the early seventeenth century as a result of investigations of various games of chance. Since then many leading mathematicians and scientists made contributions to the theory of probability. However, despite its long and active history, probability theory was not axiomatized until the twenties and thirties of this century. This axiomatic development, called modern probability theory, was now able to make the concepts of probability precise and place them on a firm mathematical foundation.

The importance of probability has increased enormously in recent years. Today the notions of probability and its twin subject statistics appear in almost every discipline, e.g. physics, chemistry, biology, medicine, psychology, sociology, political science, education, economics, business, operations research, and all fields of engineering.

This book is designed for an introductory course in probability with high school algebra as the only prerequisite. It can serve as a text for such a course, or as a supplement to all current comparable texts. The book should also prove useful as a supplement to texts and courses in statistics. Furthermore, as the book is complete and self-contained it can easily be used for self-study.

The book begins with a chapter on sets and their operations, and follows with a chapter on permutations, combinations and other techniques of counting. Next is a chapter on probability spaces and then a chapter on conditional probability and independence. The fifth and main chapter is on random variables. Here we define expectation, variance and standard deviation, and prove Tchebycheff's inequality and the law of large numbers. Although calculus is not a prerequisite, both discrete and continuous random variables are considered. We follow with a separate chapter on the binomial, normal and Poisson distributions. Here the central limit theorem is given in the context of the normal approximation to the binomial distribution. The seventh and last chapter offers a thorough elementary treatment of Markov chains with applications.

Each chapter begins with clear statements of pertinent definitions, principles and theorems together with illustrative and other descriptive material. This is followed by graded sets of solved and supplementary problems. The solved problems serve to illustrate and amplify the theory, bring into sharp focus those fine points without which the student continually feels himself on unsafe ground, and provide the repetition of basic principles so vital to effective learning. Proofs of most of the theorems are included among the solved problems. The supplementary problems serve as a complete review of the material of each chapter.

I wish to thank Dr. Martin Silverstein for his invaluable suggestions and critical review of the manuscript. I also wish to express my appreciation to Daniel Schaum and Nicola Monti for their excellent cooperation.

SEYMOUR LIPSCHUTZ

Temple University
November, 1968

CONTENTS

Chapter 1

Set Theory

INTRODUCTION

This chapter treats some of the elementary ideas and concepts of set theory which are necessary for a modern introduction to probability theory.

SETS, ELEMENTS

Any well defined list or collection of objects is called a *set*; the objects comprising the set are called its *elements* or *members*. We write

$$p \in A \quad \text{if } p \text{ is an element in the set } A$$

If every element of A also belongs to a set B, i.e. if $p \in A$ implies $p \in B$, then A is called a *subset* of B or is said to be *contained* in B; this is denoted by

$$A \subset B \quad \text{or} \quad B \supset A$$

Two sets are *equal* if each is contained in the other; that is,

$$A = B \quad \text{if and only if} \quad A \subset B \text{ and } B \subset A$$

The negations of $p \in A$, $A \subset B$ and $A = B$ are written $p \notin A$, $A \not\subset B$ and $A \neq B$ respectively.

We specify a particular set by either listing its elements or by stating properties which characterize the elements of the set. For example,

$$A = \{1, 3, 5, 7, 9\}$$

means A is the set consisting of the numbers 1, 3, 5, 7 and 9; and

$$B = \{x : x \text{ is a prime number}, \ x < 15\}$$

means that B is the set of prime numbers less than 15.

Unless otherwise stated, all sets under investigation are assumed to be subsets of some fixed set called the *universal set* and denoted (in this chapter) by U. We also use $\emptyset$ to denote the *empty* or *null* set, i.e. the set which contains no elements; this set is regarded as a subset of every other set. Thus for any set A, we have $\emptyset \subset A \subset U$.

Example 1.1: The sets A and B above can also be written as

$$A = \{x : x \text{ is an odd number}, \ x < 10\} \quad \text{and} \quad B = \{2, 3, 5, 7, 11, 13\}$$

Observe that $9 \in A$ but $9 \notin B$, and $11 \in B$ but $11 \notin A$; whereas $3 \in A$ and $3 \in B$, and $6 \notin A$ and $6 \notin B$.

Example 1.2: We use the following special symbols:

$\mathbf{N}$ = the set of positive integers: 1, 2, 3, ...

$\mathbf{Z}$ = the set of integers: ..., −2, −1, 0, 1, 2, ...

$\mathbf{R}$ = the set of real numbers.

Thus we have $\mathbf{N} \subset \mathbf{Z} \subset \mathbf{R}$.

Example 1.3: *Intervals* on the real line, defined below, appear very often in mathematics. Here a and b are real numbers with $a < b$.

Open interval from a to b $= (a, b) = \{x : a < x < b\}$

Closed interval from a to b $= [a, b] = \{x : a \leq x \leq b\}$

Open-closed interval from a to b $= (a, b] = \{x : a < x \leq b\}$

Closed-open interval from a to b $= [a, b) = \{x : a \leq x < b\}$

The open-closed and closed-open intervals are also called *half-open* intervals.

Example 1.4: In human population studies, the universal set consists of all the people in the world.

Example 1.5: Let $C = \{x : x^2 = 4, x \text{ is odd}\}$. Then $C = \emptyset$; that is, C is the empty set.

The following theorem applies.

Theorem 1.1: Let A, B and C be any sets. Then: (i) $A \subset A$; (ii) if $A \subset B$ and $B \subset A$ then $A = B$; and (iii) if $A \subset B$ and $B \subset C$ then $A \subset C$.

We emphasize that $A \subset B$ does not exclude the possibility that $A = B$. However, if $A \subset B$ but $A \neq B$, then we say that A is a *proper subset* of B. (Some authors use the symbol $\subseteq$ for a subset and the symbol $\subset$ only for a proper subset.)

SET OPERATIONS

Let A and B be arbitrary sets. The *union* of A and B, denoted by $A \cup B$, is the set of elements which belong to A or to B:

$$A \cup B = \{x : x \in A \text{ or } x \in B\}$$

Here "or" is used in the sense of and/or.

The *intersection* of A and B, denoted by $A \cap B$, is the set of elements which belong to both A and B:

$$A \cap B = \{x : x \in A \text{ and } x \in B\}$$

If $A \cap B = \emptyset$, that is, if A and B do not have any elements in common, then A and B are said to be *disjoint*.

The *difference* of A and B or the *relative complement* of B with respect to A, denoted by $A \setminus B$, is the set of elements which belong to A but not to B:

$$A \setminus B = \{x : x \in A, x \notin B\}$$

Observe that $A \setminus B$ and B are disjoint, i.e. $(A \setminus B) \cap B = \emptyset$.

The *absolute complement* or, simply, *complement* of A, denoted by A^c, is the set of elements which do not belong to A:

$$A^c = \{x : x \in U, x \notin A\}$$

That is, A^c is the difference of the universal set U and A.

Example 1.6: The following diagrams, called Venn diagrams, illustrate the above set operations. Here sets are represented by simple plane areas and U, the universal set, by the area in the entire rectangle.

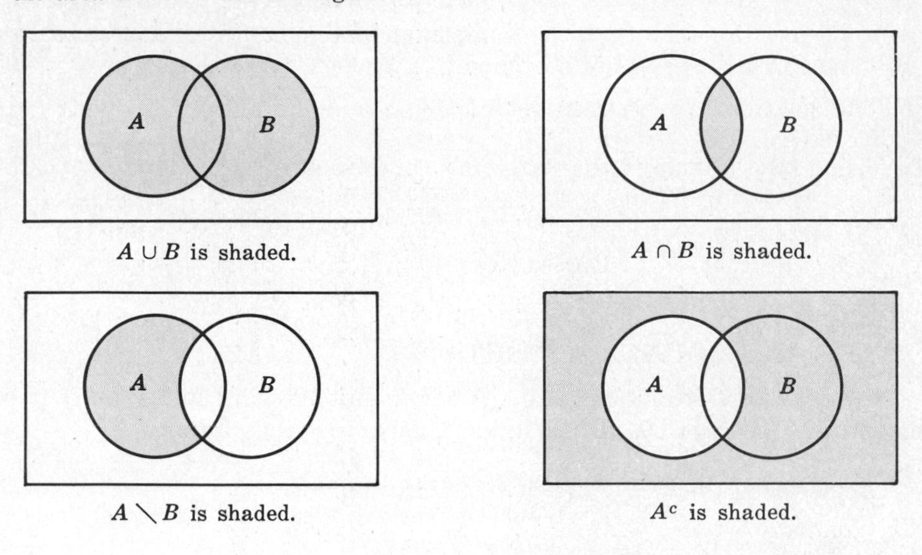

$A \cup B$ is shaded.

$A \cap B$ is shaded.

$A \setminus B$ is shaded.

A^c is shaded.

Example 1.7: Let $A = \{1,2,3,4\}$ and $B = \{3,4,5,6\}$ where $U = \{1,2,3,\ldots\}$. Then

$$A \cup B = \{1, 2, 3, 4, 5, 6\} \qquad A \cap B = \{3, 4\}$$
$$A \setminus B = \{1, 2\} \qquad A^c = \{5, 6, 7, \ldots\}$$

Sets under the above operations satisfy various laws or identities which are listed in the table below (Table 1). In fact, we state

Theorem 1.2: Sets satisfy the laws in Table 1.

LAWS OF THE ALGEBRA OF SETS	
Idempotent Laws	
1a. $A \cup A = A$	1b. $A \cap A = A$
Associative Laws	
2a. $(A \cup B) \cup C = A \cup (B \cup C)$	2b. $(A \cap B) \cap C = A \cap (B \cap C)$
Commutative Laws	
3a. $A \cup B = B \cup A$	3b. $A \cap B = B \cap A$
Distributive Laws	
4a. $A \cup (B \cap C) = (A \cup B) \cap (A \cup C)$	4b. $A \cap (B \cup C) = (A \cap B) \cup (A \cap C)$
Identity Laws	
5a. $A \cup \emptyset = A$	5b. $A \cap U = A$
6a. $A \cup U = U$	6b. $A \cap \emptyset = \emptyset$
Complement Laws	
7a. $A \cup A^c = U$	7b. $A \cap A^c = \emptyset$
8a. $(A^c)^c = A$	8b. $U^c = \emptyset$, $\emptyset^c = U$
De Morgan's Laws	
9a. $(A \cup B)^c = A^c \cap B^c$	9b. $(A \cap B)^c = A^c \cup B^c$

Table 1

Remark: Each of the above laws follows from an analogous logical law. For example,

$$A \cap B = \{x : x \in A \text{ and } x \in B\} = \{x : x \in B \text{ and } x \in A\} = B \cap A$$

Here we use the fact that the composite statement "p and q", written $p \wedge q$, is logically equivalent to the composite statement "q and p", i.e. $q \wedge p$.

The relationship between set inclusion and the above set operations follows:

Theorem 1.3: Each of the following conditions is equivalent to $A \subset B$:

(i) $A \cap B = A$ (iii) $B^c \subset A^c$ (v) $B \cup A^c = U$

(ii) $A \cup B = B$ (iv) $A \cap B^c = \emptyset$

FINITE AND COUNTABLE SETS

Sets can be finite or infinite. A set is finite if it is empty or if it consists of exactly n elements where n is a positive integer; otherwise it is infinite.

Example 1.8: Let M be the set of the days of the week; that is,

$$M = \{\text{Monday, Tuesday, Wednesday, Thursday, Friday, Saturday, Sunday}\}$$

Then M is finite.

Example 1.9: Let $P = \{x : x \text{ is a river on the earth}\}$. Although it may be difficult to count the number of rivers on the earth, P is a finite set.

Example 1.10: Let Y be the set of (positive) even integers, i.e. $Y = \{2, 4, 6, \ldots\}$. Then Y is an infinite set.

Example 1.11: Let I be the *unit interval* of real numbers, i.e. $I = \{x : 0 \leq x \leq 1\}$. Then I is also an infinite set.

A set is *countable* if it is finite or if its elements can be arranged in the form of a sequence, in which case it is said to be *countably infinite*; otherwise the set is *uncountable.* The set in Example 1.10 is countably infinite, whereas it can be shown that the set in Example 1.11 is uncountable.

PRODUCT SETS

Let A and B be two sets. The *product set* of A and B, denoted by $A \times B$, consists of all ordered pairs (a, b) where $a \in A$ and $b \in B$:

$$A \times B = \{(a, b) : a \in A, \; b \in B\}$$

The product of a set with itself, say $A \times A$, is denoted by A^2.

Example 1.12: The reader is familiar with the cartesian plane $\mathbf{R}^2 = \mathbf{R} \times \mathbf{R}$ as shown below. Here each point P represents an ordered pair (a, b) of real numbers, and vice versa.

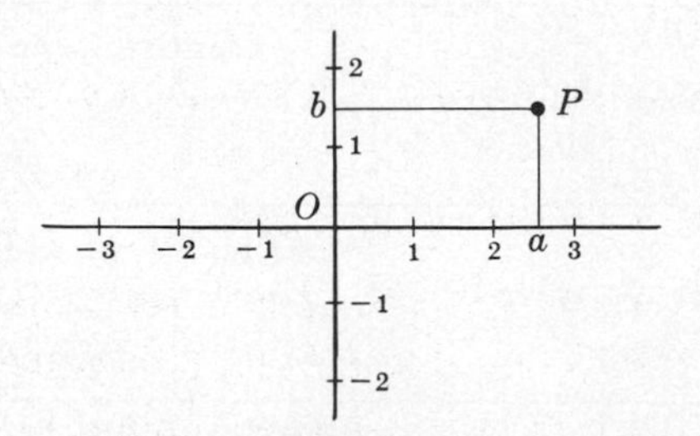

Example 1.13: Let $A = \{1, 2, 3\}$ and $B = \{a, b\}$. Then

$$A \times B = \{(1, a), (1, b), (2, a), (2, b), (3, a), (3, b)\}$$

The concept of product set is extended to any finite number of sets in a natural way. The product set of the sets $A_1, A_2, \ldots, A_m$, written $A_1 \times A_2 \times \cdots \times A_m$, is the set of all ordered m-tuples $(a_1, a_2, \ldots, a_m)$ where $a_i \in A_i$ for each i.

CLASSES OF SETS

Frequently the members of a set are sets themselves. For example, each line in a set of lines is a set of points. To help clarify these situations, we usually use the word *class* or *family* for such a set. The words subclass and subfamily have meanings analogous to subset.

Example 1.14: The members of the class $\{\{2,3\}, \{2\}, \{5,6\}\}$ are the sets $\{2,3\}$, $\{2\}$ and $\{5,6\}$.

Example 1.15: Consider any set A. The *power set* of A, denoted by $\mathcal{P}(A)$, is the class of all subsets of A. In particular, if $A = \{a, b, c\}$, then

$$\mathcal{P}(A) = \{A, \{a,b\}, \{a,c\}, \{b,c\}, \{a\}, \{b\}, \{c\}, \emptyset\}$$

In general, if A is finite and has n elements, then $\mathcal{P}(A)$ will have 2^n elements.

A *partition* of a set X is a subdivision of X into nonempty subsets which are disjoint and whose union is X, i.e. is a class of nonempty subsets of X such that each $a \in X$ belongs to a unique subset. The subsets in a partition are called *cells*.

Example 1.16: Consider the following classes of subsets of $X = \{1, 2, \ldots, 8, 9\}$:

(i) $[\{1,3,5\}, \{2,6\}, \{4,8,9\}]$

(ii) $[\{1,3,5\}, \{2,4,6,8\}, \{5,7,9\}]$

(iii) $[\{1,3,5\}, \{2,4,6,8\}, \{7,9\}]$

Then (i) is not a partition of X since $7 \in X$ but 7 does not belong to any of the cells. Furthermore, (ii) is not a partition of X since $5 \in X$ and 5 belongs to both $\{1,3,5\}$ and $\{5,7,9\}$. On the other hand, (iii) is a partition of X since each element of X belongs to exactly one cell.

When we speak of an *indexed class of sets* $\{A_i : i \in I\}$ or simply $\{A_i\}$, we mean that there is a set A_i assigned to each element $i \in I$. The set I is called the *indexing set* and the sets A_i are said to be indexed by I. When the indexing set is the set $\mathbf{N}$ of positive integers, the indexed class $\{A_1, A_2, \ldots\}$ is called a *sequence* of sets. By the *union* of these A_i, denoted by $\cup_{i \in I} A_i$ (or simply $\cup_i A_i$), we mean the set of elements each belonging to at least one of the A_i; and by the *intersection* of the A_i, denoted by $\cap_{i \in I} A_i$ (or simply $\cap_i A_i$), we mean the set of elements each belonging to every A_i. We also write

$$\cup_{i=1}^{\infty} A_i = A_1 \cup A_2 \cup \cdots \quad \text{and} \quad \cap_{i=1}^{\infty} A_i = A_1 \cap A_2 \cap \cdots$$

for the union and intersection, respectively, of a sequence of sets.

Definition: A nonempty class $\mathcal{A}$ of subsets of U is called an *algebra* (σ-*algebra*) of sets if:

(i) the complement of any set in $\mathcal{A}$ belongs to $\mathcal{A}$; and

(ii) the union of any finite (countable) number of sets in $\mathcal{A}$ belongs to $\mathcal{A}$;

that is, if $\mathcal{A}$ is closed under complements and finite (countable) unions.

It is simple to show (Problem 1.30) that an algebra (σ-algebra) of sets contains U and $\emptyset$ and is also closed under finite (countable) intersections.

Solved Problems

SETS, ELEMENTS, SUBSETS

1.1. Let $A = \{x : 3x = 6\}$. Does $A = 2$?

A is the set which consists of the single element 2, that is, $A = \{2\}$. The number 2 belongs to A; it does not equal A. There is a basic difference between an element p and the singleton set $\{p\}$.

1.2. Which of these sets are equal: $\{r, s, t\}$, $\{t, s, r\}$, $\{s, r, t\}$, $\{t, r, s\}$?

They are all equal. Order does not change a set.

1.3. Determine whether or not each set is the null set:
(i) $X = \{x : x^2 = 9,\ 2x = 4\}$, (ii) $Y = \{x : x \neq x\}$, (iii) $Z = \{x : x + 8 = 8\}$.

(i) There is no number which satisfies both $x^2 = 9$ and $2x = 4$; hence X is empty, i.e. $X = \emptyset$.

(ii) We interpret "=" to mean "is identical with" and so Y is also empty. In fact, some texts define the empty set as follows: $\emptyset \equiv \{x : x \neq x\}$.

(iii) The number zero satisfies $x + 8 = 8$; hence $Z = \{0\}$. Accordingly, Z is not the empty set since it contains 0. That is, $Z \neq \emptyset$.

1.4. Prove that $A = \{2, 3, 4, 5\}$ is not a subset of $B = \{x : x \text{ is even}\}$.

It is necessary to show that at least one element in A does not belong to B. Now $3 \in A$ and, since B consists of even numbers, $3 \notin B$; hence A is not a subset of B.

1.5. Let $V = \{d\}$, $W = \{c, d\}$, $X = \{a, b, c\}$, $Y = \{a, b\}$ and $Z = \{a, b, d\}$. Determine whether each statement is true or false:
(i) $Y \subset X$, (ii) $W \neq Z$, (iii) $Z \supset V$, (iv) $V \subset X$, (v) $X = W$, (vi) $W \subset Y$.

(i) Since each element in Y is a member of X, $Y \subset X$ is true.

(ii) Now $a \in Z$ but $a \notin W$; hence $W \neq Z$ is true.

(iii) The only element in V is d and it also belongs to Z; hence $Z \supset V$ is true.

(iv) V is not a subset of X since $d \in V$ but $d \notin X$; hence $V \subset X$ is false.

(v) Now $a \in X$ but $a \notin W$; hence $X = W$ is false.

(vi) W is not a subset of Y since $c \in W$ but $c \notin Y$; hence $W \subset Y$ is false.

1.6. Prove: If A is a subset of the empty set $\emptyset$, then $A = \emptyset$.

The null set $\emptyset$ is a subset of every set; in particular, $\emptyset \subset A$. But, by hypothesis, $A \subset \emptyset$; hence $A = \emptyset$.

1.7. Prove Theorem 1.1(iii): If $A \subset B$ and $B \subset C$, then $A \subset C$.

We must show that each element in A also belongs to C. Let $x \in A$. Now $A \subset B$ implies $x \in B$. But $B \subset C$; hence $x \in C$. We have shown that $x \in A$ implies $x \in C$, that is, that $A \subset C$.

1.8. Which of the following sets are finite?

(i) The months of the year.

(ii) $\{1, 2, 3, \ldots, 99, 100\}$.

(iii) The number of people living on the earth.

(iv) The set **Q** of rational numbers.

(v) The set **R** of real numbers.

The first three sets are finite; the last two are infinite. (It can be shown that **Q** is countable but **R** is uncountable.)

1.9. Consider the following sets of figures in the Euclidean plane:

$$A = \{x : x \text{ is a quadrilateral}\} \qquad C = \{x : x \text{ is a rhombus}\}$$
$$B = \{x : x \text{ is a rectangle}\} \qquad D = \{x : x \text{ is a square}\}$$

Determine which sets are proper subsets of any of the others.

Since a square has 4 right angles it is a rectangle, since it has 4 equal sides it is a rhombus, and since it has 4 sides it is a quadrilateral. Thus

$$D \subset A, \quad D \subset B \quad \text{and} \quad D \subset C$$

that is, D is a subset of the other three. Also, since there are examples of rectangles, rhombuses and quadrilaterals which are not squares, D is a proper subset of the other three.

In a similar manner we see that B is a proper subset of A and C is a proper subset of A. There are no other relations among the sets.

1.10. Determine which of the following sets are equal: $\emptyset$, $\{0\}$, $\{\emptyset\}$.

Each is different from the other. The set $\{0\}$ contains one element, the number zero. The set $\emptyset$ contains no elements; it is the empty set. The set $\{\emptyset\}$ also contains one element, the null set.

SET OPERATIONS

1.11. Let $U = \{1, 2, \ldots, 8, 9\}$, $A = \{1, 2, 3, 4\}$, $B = \{2, 4, 6, 8\}$ and $C = \{3, 4, 5, 6\}$. Find: (i) A^c, (ii) $A \cap C$, (iii) $(A \cap C)^c$, (iv) $A \cup B$, (v) $B \setminus C$.

(i) A^c consists of the elements in U that are not in A; hence $A^c = \{5, 6, 7, 8, 9\}$.

(ii) $A \cap C$ consists of the elements in both A and C; hence $A \cap C = \{3, 4\}$.

(iii) $(A \cap C)^c$ consists of the elements in U that are not in $A \cap C$. Now by (ii), $A \cap C = \{3, 4\}$ and so $(A \cap C)^c = \{1, 2, 5, 6, 7, 8, 9\}$.

(iv) $A \cup B$ consists of the elements in A or B (or both): hence $A \cup B = \{1, 2, 3, 4, 6, 8\}$.

(v) $B \setminus C$ consists of the elements in B which are not in C; hence $B \setminus C = \{2, 8\}$.

1.12. Let $U = \{a, b, c, d, e\}$, $A = \{a, b, d\}$ and $B = \{b, d, e\}$. Find:

(i) $A \cup B$ (iii) B^c (v) $A^c \cap B$ (vii) $A^c \cap B^c$ (ix) $(A \cap B)^c$

(ii) $B \cap A$ (iv) $B \setminus A$ (vi) $A \cup B^c$ (viii) $B^c \setminus A^c$ (x) $(A \cup B)^c$

(i) The union of A and B consists of the elements in A or in B (or both); hence $A \cup B = \{a, b, d, e\}$.

(ii) The intersection of A and B consists of those elements which belong to both A and B; hence $A \cap B = \{b, d\}$.

(iii) The complement of B consists of the letters in U but not in B; hence $B^c = \{a, c\}$.

(iv) The difference $B \setminus A$ consists of the elements of B which do not belong to A; hence $B \setminus A = \{e\}$.

(v) $A^c = \{c, e\}$ and $B = \{b, d, e\}$; then $A^c \cap B = \{e\}$.

(vi) $A = \{a, b, d\}$ and $B^c = \{a, c\}$; then $A \cup B^c = \{a, b, c, d\}$.

(vii) and (viii). $A^c = \{c, e\}$ and $B^c = \{a, c\}$; then

$$A^c \cap B^c = \{c\} \qquad \text{and} \qquad B^c \setminus A^c = \{a\}$$

(ix) From (ii), $A \cap B = \{b, d\}$; hence $(A \cap B)^c = \{a, c, e\}$.

(x) From (i), $A \cup B = \{a, b, d, e\}$; hence $(A \cup B)^c = \{c\}$.

1.13. In the Venn diagram below, shade: (i) B^c, (ii) $(A \cup B)^c$, (iii) $(B \setminus A)^c$, (iv) $A^c \cap B^c$.

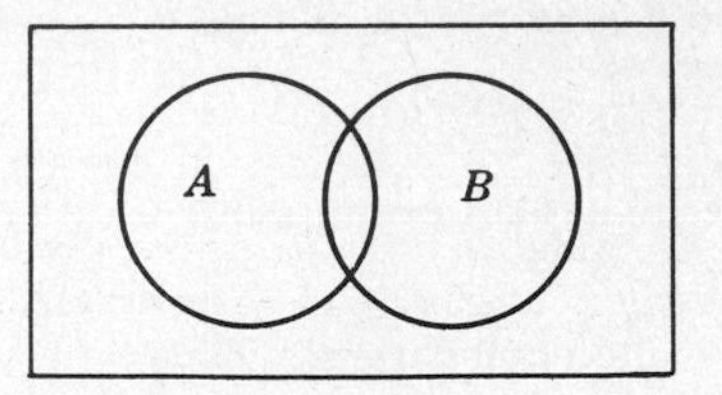

(i) B^c consists of the elements which do not belong to B; hence shade the area outside B as follows:

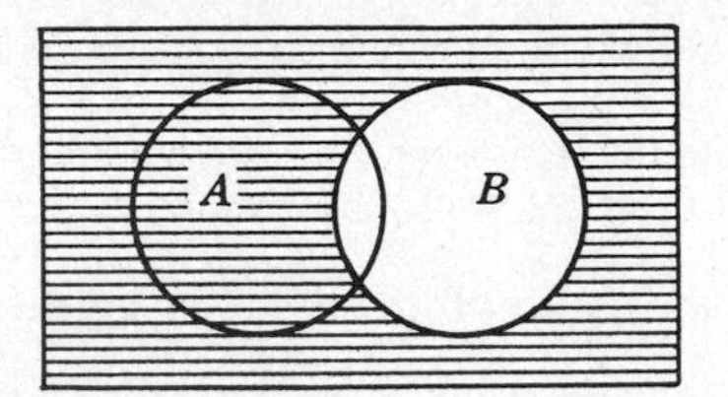

B^c is shaded.

(ii) First shade $A \cup B$; then $(A \cup B)^c$ is the area outside $A \cup B$:

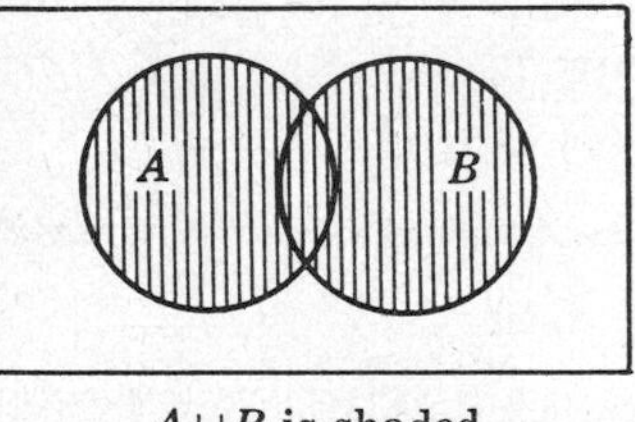

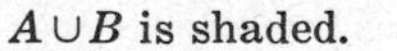

$A \cup B$ is shaded.

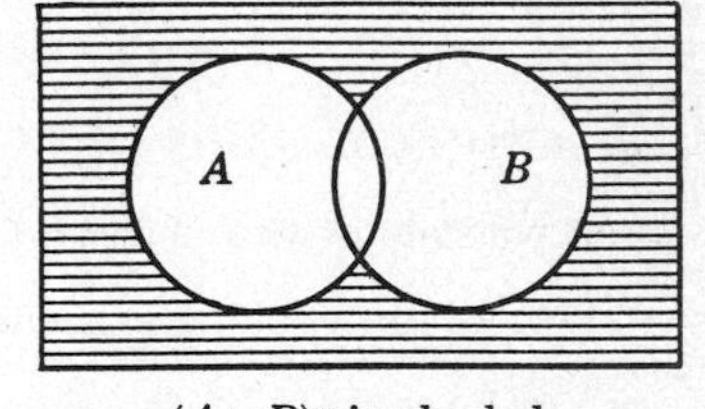

$(A \cup B)^c$ is shaded.

(iii) First shade $B \setminus A$, the area in B which does not lie in A; then $(B \setminus A)^c$ is the area outside $B \setminus A$:

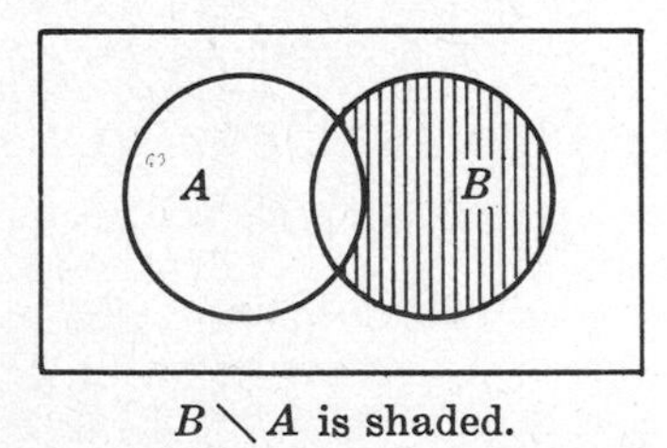

$B \setminus A$ is shaded.

$(B \setminus A)^c$ is shaded.

(iv) First shade A^c, the area outside of A, with strokes slanting upward to the right (////), and then shade B^c with strokes slanting downward to the right (\\\\); then $A^c \cap B^c$ is the cross-hatched area:

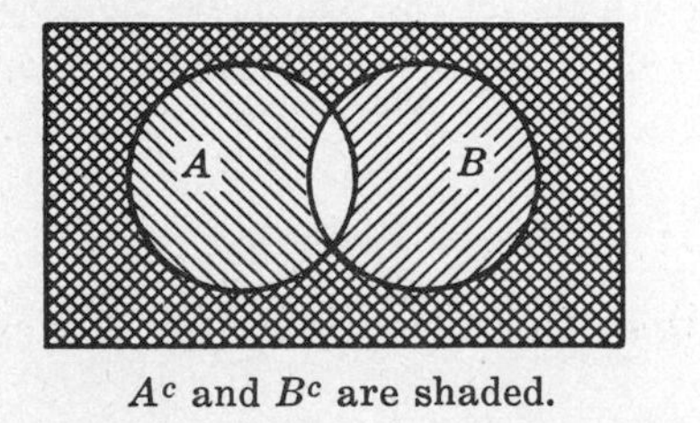

A^c and B^c are shaded.

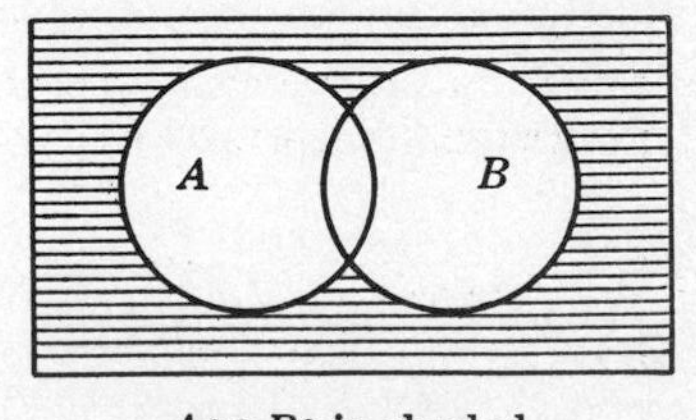

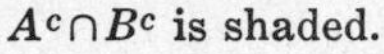

$A^c \cap B^c$ is shaded.

Observe that $(A \cup B)^c = A^c \cap B^c$, as expected by De Morgan's law.

1.14. Prove: $B \setminus A = B \cap A^c$. Thus the set operation of difference can be written in terms of the operations of intersection and complementation.

$$B \setminus A \quad = \quad \{x : x \in B,\ x \notin A\} \quad = \quad \{x : x \in B,\ x \in A^c\} \quad = \quad B \cap A^c$$

1.15. Prove: For any sets A and B, $A \cap B \subset A \subset A \cup B$.

Let $x \in A \cap B$; then $x \in A$ and $x \in B$. In particular, $x \in A$. Since $x \in A \cap B$ implies $x \in A$, $A \cap B \subset A$. Furthermore if $x \in A$, then $x \in A$ or $x \in B$, i.e. $x \in A \cup B$. Hence $A \subset A \cup B$. In other words, $A \cap B \subset A \subset A \cup B$.

1.16. Prove Theorem 1.3(i): $A \subset B$ if and only if $A \cap B = A$.

Suppose $A \subset B$. Let $x \in A$; then by hypothesis, $x \in B$. Hence $x \in A$ and $x \in B$, i.e. $x \in A \cap B$. Accordingly, $A \subset A \cap B$. On the other hand, it is always true (Problem 1.15) that $A \cap B \subset A$. Thus $A \cap B = A$.

Now suppose that $A \cap B = A$. Then in particular, $A \subset A \cap B$. But it is always true that $A \cap B \subset B$. Thus $A \subset A \cap B \subset B$ and so, by Theorem 1.1, $A \subset B$.

PRODUCT SETS

1.17. Let M = {Tom, Marc, Erik} and W = {Audrey, Betty}. Find $M \times W$.

$M \times W$ consists of all ordered pairs (a, b) where $a \in M$ and $b \in W$. Hence

$M \times W$ = {(Tom, Audrey), (Tom, Betty), (Marc, Audrey), (Marc, Betty), (Erik, Audrey), (Erik, Betty)}

1.18. Let $A = \{1, 2, 3\}$, $B = \{2, 4\}$ and $C = \{3, 4, 5\}$. Find $A \times B \times C$.

A convenient method of finding $A \times B \times C$ is through the so-called "tree diagram" shown below:

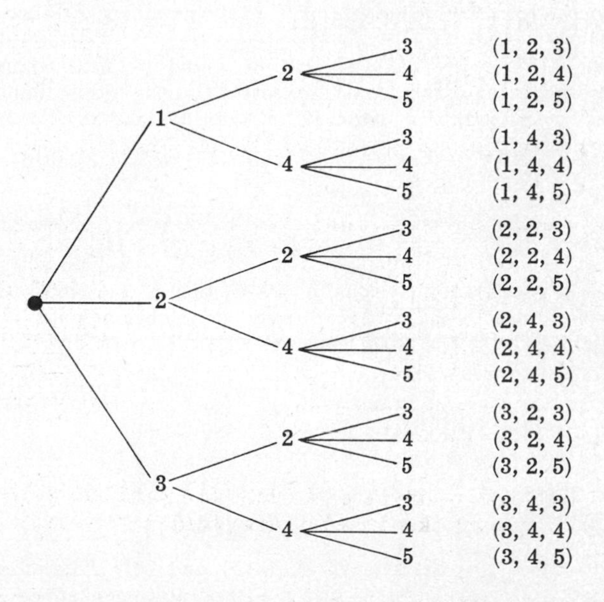

The "tree" is constructed from the left to the right. $A \times B \times C$ consists of the ordered triples listed to the right of the "tree".

1.19. Let $A = \{a, b\}$, $B = \{2, 3\}$ and $C = \{3, 4\}$. Find:
(i) $A \times (B \cup C)$, (ii) $(A \times B) \cup (A \times C)$, (iii) $A \times (B \cap C)$, (iv) $(A \times B) \cap (A \times C)$.

(i) First compute $B \cup C = \{2, 3, 4\}$. Then

$$A \times (B \cup C) \quad = \quad \{(a, 2), (a, 3), (a, 4), (b, 2), (b, 3), (b, 4)\}$$

(ii) First find $A \times B$ and $A \times C$:

$$A \times B = \{(a,2), (a,3), (b,2), (b,3)\}$$
$$A \times C = \{(a,3), (a,4), (b,3), (b,4)\}$$

Then compute the union of the two sets:

$$(A \times B) \cup (A \times C) = \{(a,2), (a,3), (b,2), (b,3), (a,4), (b,4)\}$$

Observe from (i) and (ii) that

$$A \times (B \cup C) = (A \times B) \cup (A \times C)$$

(iii) First compute $B \cap C = \{3\}$. Then

$$A \times (B \cap C) = \{(a,3), (b,3)\}$$

(iv) Now $A \times B$ and $A \times C$ were computed above. The intersection of $A \times B$ and $A \times C$ consists of those ordered pairs which belong to both sets:

$$(A \times B) \cap (A \times C) = \{(a,3), (b,3)\}$$

Observe from (iii) and (iv) that

$$A \times (B \cap C) = (A \times B) \cap (A \times C)$$

1.20. Prove: $A \times (B \cap C) = (A \times B) \cap (A \times C)$.

$$\begin{aligned} A \times (B \cap C) &= \{(x,y) : x \in A,\ y \in B \cap C\} \\ &= \{(x,y) : x \in A,\ y \in B,\ y \in C\} \\ &= \{(x,y) : (x,y) \in A \times B,\ (x,y) \in A \times C\} \\ &= (A \times B) \cap (A \times C) \end{aligned}$$

1.21. Let $S = \{a, b\}$, $W = \{1,2,3,4,5,6\}$ and $V = \{3,5,7,9\}$. Find $(S \times W) \cap (S \times V)$.

The product set $(S \times W) \cap (S \times V)$ can be found by first computing $S \times W$ and $S \times V$, and then computing the intersection of these sets. On the other hand, by the preceding problem, $(S \times W) \cap (S \times V) = S \times (W \cap V)$. Now $W \cap V = \{3,5\}$, and so

$$(S \times W) \cap (S \times V) = S \times (W \cap V) = \{(a,3), (a,5), (b,3), (b,5)\}$$

1.22. Prove: Let $A \subset B$ and $C \subset D$; then $(A \times C) \subset (B \times D)$.

Let (x,y) be any arbitrary element in $A \times C$; then $x \in A$ and $y \in C$. By hypothesis, $A \subset B$ and $C \subset D$; hence $x \in B$ and $y \in D$. Accordingly (x,y) belongs to $B \times D$. We have shown that $(x,y) \in A \times C$ implies $(x,y) \in B \times D$; hence $(A \times C) \subset (B \times D)$.

CLASSES OF SETS

1.23. Consider the class $A = \{\{2,3\}, \{4,5\}, \{6\}\}$. Which statements are incorrect and why? (i) $\{4,5\} \subset A$, (ii) $\{4,5\} \in A$, (iii) $\{\{4,5\}\} \subset A$.

The members of A are the sets $\{2,3\}$, $\{4,5\}$ and $\{6\}$. Therefore (ii) is correct but (i) is an incorrect statement. Moreover, (iii) is also a correct statement since the set consisting of the single element $\{4,5\}$ is a subclass of A.

1.24. Find the power set $\mathcal{P}(S)$ of the set $S = \{1,2,3\}$.

The power set $\mathcal{P}(S)$ of S is the class of all subsets of S; these are $\{1,2,3\}$, $\{1,2\}$, $\{1,3\}$, $\{2,3\}$, $\{1\}$, $\{2\}$, $\{3\}$ and the empty set $\emptyset$. Hence

$$\mathcal{P}(S) = \{S, \{1,3\}, \{2,3\}, \{1,2\}, \{1\}, \{2\}, \{3\}, \emptyset\}$$

Note that there are $2^3 = 8$ subsets of S.

1.25. Let $X = \{a, b, c, d, e, f, g\}$, and let:

(i) $A_1 = \{a, c, e\}$, $A_2 = \{b\}$, $A_3 = \{d, g\}$;

(ii) $B_1 = \{a, e, g\}$, $B_2 = \{c, d\}$, $B_3 = \{b, e, f\}$;

(iii) $C_1 = \{a, b, e, g\}$, $C_2 = \{c\}$, $C_3 = \{d, f\}$;

(iv) $D_1 = \{a, b, c, d, e, f, g\}$.

Which of $\{A_1, A_2, A_3\}$, $\{B_1, B_2, B_3\}$, $\{C_1, C_2, C_3\}$, $\{D_1\}$ are partitions of X?

(i) $\{A_1, A_2, A_3\}$ is not a partition of X since $f \in X$ but f does not belong to either A_1, A_2, or A_3.

(ii) $\{B_1, B_2, B_3\}$ is not a partition of X since $e \in X$ belongs to both B_1 and B_3.

(iii) $\{C_1, C_2, C_3\}$ is a partition of X since each element in X belongs to exactly one cell, i.e. $X = C_1 \cup C_2 \cup C_3$ and the sets are pairwise disjoint.

(iv) $\{D_1\}$ is a partition of X.

1.26. Find all the partitions of $X = \{a, b, c, d\}$.

Note first that each partition of X contains either 1, 2, 3, or 4 distinct sets. The partitions are as follows:

(1) $[\{a, b, c, d\}]$

(2) $[\{a\}, \{b, c, d\}]$, $[\{b\}, \{a, c, d\}]$, $[\{c\}, \{a, b, d\}]$, $[\{d\}, \{a, b, c\}]$,
$[\{a, b\}, \{c, d\}]$, $[\{a, c\}, \{b, d\}]$, $[\{a, d\}, \{b, c\}]$

(3) $[\{a\}, \{b\}, \{c, d\}]$, $[\{a\}, \{c\}, \{b, d\}]$, $[\{a\}, \{d\}, \{b, c\}]$,
$[\{b\}, \{c\}, \{a, d\}]$, $[\{b\}, \{d\}, \{a, c\}]$, $[\{c\}, \{d\}, \{a, b\}]$

(4) $[\{a\}, \{b\}, \{c\}, \{d\}]$

There are fifteen different partitions of X.

1.27. Let $\mathbf{N}$ be the set of positive integers and, for each $n \in \mathbf{N}$, let

$$A_n = \{x : x \text{ is a multiple of } n\} = \{n, 2n, 3n, \ldots\}$$

Find (i) $A_3 \cap A_5$, (ii) $A_4 \cap A_6$, (iii) $\cup_{i \in P} A_i$, where P is the set of prime numbers, 2, 3, 5, 7, 11, ...

(i) Those numbers which are multiples of both 3 and 5 are the multiples of 15; hence $A_3 \cap A_5 = A_{15}$.

(ii) The multiples of 12 and no other numbers belong to both A_4 and A_6; hence $A_4 \cap A_6 = A_{12}$.

(iii) Every positive integer except 1 is a multiple of at least one prime number; hence

$$\cup_{i \in P} A_i = \{2, 3, 4, \ldots\} = \mathbf{N} \setminus \{1\}$$

1.28. Prove: Let $\{A_i : i \in I\}$ be an indexed class of sets and let $i_0 \in I$. Then

$$\cap_{i \in I} A_i \subset A_{i_0} \subset \cup_{i \in I} A_i$$

Let $x \in \cap_{i \in I} A_i$; then $x \in A_i$ for every $i \in I$. In particular, $x \in A_{i_0}$. Hence $\cap_{i \in I} A_i \subset A_{i_0}$. Now let $y \in A_{i_0}$. Since $i_0 \in I$, $y \in \cup_{i \in I} A_i$. Hence $A_{i_0} \subset \cup_{i \in I} A_i$.

1.29. Prove (De Morgan's law): For any indexed class $\{A_i : i \in I\}$, $(\cup_i A_i)^c = \cap_i A_i^c$.

$$(\cup_i A_i)^c = \{x : x \notin \cup_i A_i\} = \{x : x \notin A_i \text{ for every } i\} = \{x : x \in A_i^c \text{ for every } i\} = \cap_i A_i^c$$

1.30. Let $\mathcal{A}$ be an algebra (σ-algebra) of subsets of U. Show that: (i) U and $\emptyset$ belong to $\mathcal{A}$; and (ii) $\mathcal{A}$ is closed under finite (countable) intersections.

Recall that $\mathcal{A}$ is closed under complements and finite (countable) unions.

(i) Since $\mathcal{A}$ is nonempty, there is a set $A \in \mathcal{A}$. Hence the complement $A^c \in \mathcal{A}$, and the union $U = A \cup A^c \in \mathcal{A}$. Also the complement $\emptyset = U^c \in \mathcal{A}$.

(ii) Let $\{A_i\}$ be a finite (countable) class of sets belonging to $\mathcal{A}$. By De Morgan's law (Problem 1.29), $(\cup_i A_i^c)^c = \cap_i A_i^{cc} = \cap_i A_1$. Hence $\cap_i A_i$ belongs to $\mathcal{A}$, as required.

Supplementary Problems

SETS, ELEMENTS, SUBSETS

1.31. Write in set notation:

(*a*) R is a subset of T.
(*b*) x is a member of Y.
(*c*) The empty set.
(*d*) M is not a subset of S.
(*e*) z does not belong to A.
(*f*) R belongs to $\mathcal{A}$.

1.32. Rewrite explicitly giving the elements in each set:

(i) $A = \{x : x^2 - x - 2 = 0\}$
(ii) $B = \{x : x \text{ is a letter in the word "follow"}\}$
(iii) $C = \{x : x^2 = 9,\ x - 3 = 5\}$
(iv) $D = \{x : x \text{ is a vowel}\}$
(v) $E = \{x : x \text{ is a digit in the number } 2324\}$

1.33. Let $A = \{1, 2, \ldots, 8, 9\}$, $B = \{2, 4, 6, 8\}$, $C = \{1, 3, 5, 7, 9\}$, $D = \{3, 4, 5\}$ and $E = \{3, 5\}$. Which sets can equal X if we are given the following information?
(i) X and B are disjoint. (ii) $X \subset D$ but $X \not\subset B$. (iii) $X \subset A$ but $X \not\subset C$. (iv) $X \subset C$ but $X \not\subset A$.

1.34. State whether each statement is true or false:

(i) $\{1, 4, 3\} = \{3, 4, 1\}$
(ii) $\{3, 1, 2\} \subset \{1, 2, 3\}$
(iii) $1 \not\subset \{1, 2\}$
(iv) $\{4\} \in \{\{4\}\}$
(v) $\{4\} \subset \{\{4\}\}$
(vi) $\emptyset \subset \{\{4\}\}$

1.35. Let $A = \{1, 0\}$. State whether or not each statement is correct:
(i) $\{0\} \in A$, (ii) $\emptyset \in A$, (iii) $\{0\} \subset A$, (iv) $0 \in A$, (v) $0 \subset A$.

1.36. State whether each set is finite or infinite:

(i) The set of lines parallel to the x axis.
(ii) The set of letters in the English alphabet.
(iii) The set of numbers which are multiples of 5.
(iv) The set of animals living on the earth.
(v) The set of numbers which are solutions of the equation $x^{27} + 26x^{18} - 17x^{11} + 7x^3 - 10 = 0$.
(vi) The set of circles through the origin $(0, 0)$.

SET OPERATIONS

1.37. Let $U = \{a, b, c, d, e, f, g\}$, $A = \{a, b, c, d, e\}$, $B = \{a, c, e, g\}$ and $C = \{b, e, f, g\}$. Find:

(i) $A \cup C$
(ii) $B \cap A$
(iii) $C \setminus B$
(iv) $B^c \cup C$
(v) $C^c \cap A$
(vi) $(A \setminus C)^c$
(vii) $(A \setminus B^c)^c$
(viii) $(A \cap A^c)^c$

1.38. In the Venn diagrams below, shade (i) $W \setminus V$, (ii) $V^c \cup W$, (iii) $V \cap W^c$, (iv) $V^c \setminus W^c$.

V W

(*a*)

V W

(*b*)

1.39. Prove: (*a*) $A \cup B = (A^c \cap B^c)^c$; (*b*) $A \setminus B = A \cap B^c$. (Thus the union and difference operations can be defined in terms of the operations of intersection and complement.)

1.40. Prove Theorem 1.3(ii): $A \subset B$ if and only if $A \cup B = B$.

1.41. Prove: If $A \cap B = \emptyset$, then $A \subset B^c$.

1.42. Prove: $A^c \setminus B^c = B \setminus A$.

1.43. Prove: $A \subset B$ implies $A \cup (B \setminus A) = B$.

1.44. (i) Prove: $A \cap (B \setminus C) = (A \cap B) \setminus (A \cap C)$.
(ii) Give an example to show that $A \cup (B \setminus C) \neq (A \cup B) \setminus (A \cup C)$.

PRODUCT SETS

1.45. Let $W = \{\text{Mark, Eric, Paul}\}$ and let $V = \{\text{Eric, David}\}$. Find:
(i) $W \times V$, (ii) $V \times W$, (iii) $V^2 = V \times V$.

1.46. Let $A = \{2, 3\}$, $B = \{1, 3, 5\}$ and $C = \{3, 4\}$. Construct the "tree diagram" of $A \times B \times C$ and then find $A \times B \times C$. (See Problem 1.18.)

1.47. Let $S = \{a, b, c\}$, $T = \{b, c, d\}$ and $W = \{a, d\}$. Construct the tree diagram of $S \times T \times W$ and then find $S \times T \times W$.

1.48. Suppose that the sets V, W and Z have 3, 4 and 5 elements respectively. Determine the number of elements in (i) $V \times W \times Z$, (ii) $Z \times V \times W$, (iii) $W \times Z \times V$.

1.49. Let $A = B \cap C$. Determine if either statement is true:
(i) $A \times A = (B \times B) \cap (C \times C)$, (ii) $A \times A = (B \times C) \cap (C \times B)$.

1.50. Prove: $A \times (B \cup C) = (A \times B) \cup (A \times C)$.

CLASSES OF SETS

1.51. Let $A_n = \{x : x \text{ is a multiple of } n\} = \{n, 2n, 3n, \ldots\}$, where $n \in \mathbf{N}$, the positive integers. Find: (i) $A_2 \cap A_7$; (ii) $A_6 \cap A_8$; (iii) $A_3 \cup A_{12}$; (iv) $A_3 \cap A_{12}$; (v) $A_s \cup A_{st}$, where $s, t \in \mathbf{N}$; (vi) $A_s \cap A_{st}$, where $s, t \in \mathbf{N}$. (vii) Prove: If $J \subset \mathbf{N}$ is infinite, then $\cap_{i \in J} A_i = \emptyset$.

1.52. Find the power set $\mathcal{P}(A)$ of $A = \{1, 2, 3, 4\}$ and the power set $\mathcal{P}(B)$ of $B = \{1, \{2, 3\}, 4\}$.

1.53. Let $W = \{1, 2, 3, 4, 5, 6\}$. Determine whether each of the following is a partition of W:

(i) $[\{1, 3, 5\}, \{2, 4\}, \{3, 6\}]$ (iii) $[\{1, 5\}, \{2\}, \{4\}, \{3, 6\}]$
(ii) $[\{1, 5\}, \{2\}, \{3, 6\}]$ (iv) $[\{1, 2, 3, 4, 5, 6\}]$

1.54. Find all partitions of $V = \{1, 2, 3\}$.

1.55. Let $[A_1, A_2, \ldots, A_m]$ and $[B_1, B_2, \ldots, B_n]$ be partitions of a set X. Show that the collection of sets
$$[A_i \cap B_j : i = 1, \ldots, m, j = 1, \ldots, n]$$
is also a partition (called the *cross partition*) of X.

1.56. Prove: For any indexed class $\{A_i : i \in I\}$ and any set B,
(*a*) $B \cup (\cap_i A_i) = \cap_i (B \cup A_i)$, (*b*) $B \cap (\cup_i A_i) = \cup_i (B \cap A_i)$.

1.57. Prove (De Morgan's law): $(\cap_i A_i)^c = \cup_i A_i^c$.

1.58. Show that each of the following is an algebra of subsets of U:
(i) $\mathcal{A} = \{\emptyset, U\}$; (ii) $\mathcal{B} = \{\emptyset, A, A^c, U\}$; (iii) $\mathcal{P}(U)$, the power set of U.

1.59. Let $\mathcal{A}$ and $\mathcal{B}$ be algebras (σ-algebras) of subsets of U. Prove that the intersection $\mathcal{A} \cap \mathcal{B}$ is also an algebra (σ-algebra) of subsets of U.

Answers to Supplementary Problems

1.31. (*a*) $R \subset T$, (*b*) $x \in Y$, (*c*) $\emptyset$, (*d*) $M \not\subset S$, (*e*) $z \notin A$, (*f*) $R \in \mathcal{A}$.

1.32. (i) $A = \{-1, 2\}$, (ii) $B = \{f, o, l, w\}$, (iii) $C = \emptyset$, (iv) $D = \{a, e, i, o, u\}$, (v) $E = \{2, 3, 4\}$.

1.33. (i) C and E, (ii) D and E, (iii) A, B and D, (iv) none.

1.34. All the statements are true except (v).

1.35. (i) incorrect, (ii) incorrect, (iii) correct, (iv) correct, (v) incorrect.

1.36. (i) infinite, (ii) finite, (iii) infinite, (iv) finite, (v) finite, (vi) infinite.

1.37.
(i) $A \cup C = U$
(ii) $B \cap A = \{a, c, e\}$
(iii) $C \setminus B = \{b, f\}$
(iv) $B^c \cup C = \{b, d, e, f, g\}$
(v) $C^c \cap A = \{a, c, d\} = C^c$
(vi) $(A \setminus C)^c = \{b, e, f, g\}$
(vii) $(A \setminus B^c)^c = \{b, d, f, g\}$
(viii) $(A \cap A^c)^c = U$

1.38. (*a*)

(*b*)

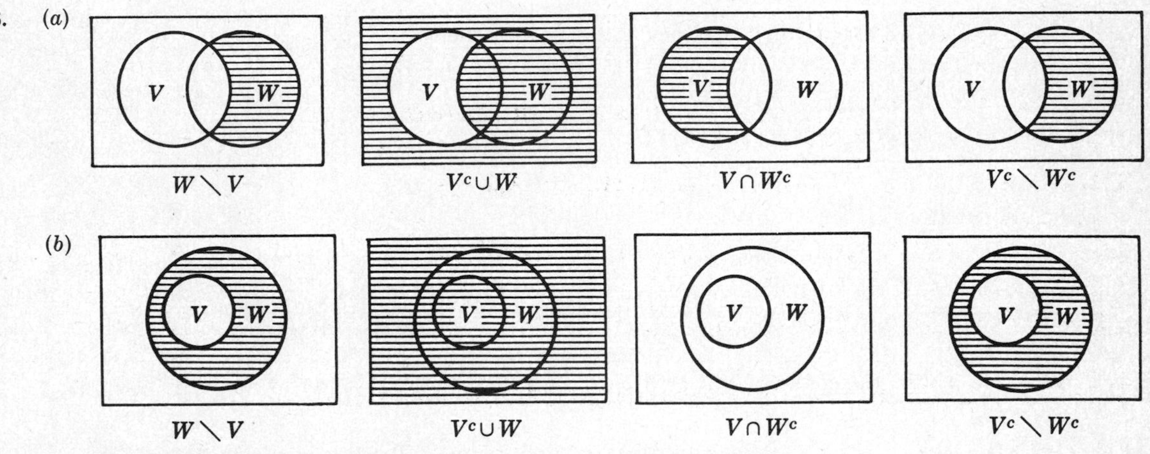

Observe that $V^c \cup W = U$ and $V \cap W^c = \emptyset$ in case (*b*) where $V \subset W$.

1.45.
(i) $W \times V$ = {(Mark, Eric), (Mark, David), (Eric, Eric), (Eric, David), (Paul, Eric), (Paul, David)}
(ii) $V \times W$ = {(Eric, Mark), (David, Mark), (Eric, Eric), (David, Eric), (Eric, Paul), (David, Paul)}
(iii) $V \times V$ = {(Eric, Eric), (Eric, David), (David, Eric), (David, David)}

1.46.

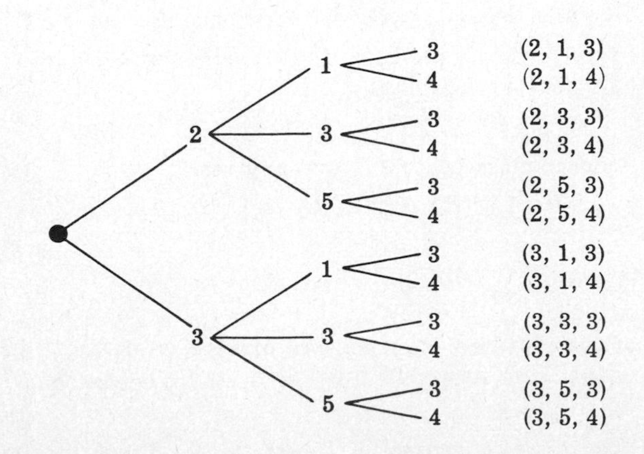

The elements of $A \times B \times C$ are the ordered triplets to the right of the tree diagram above.

1.47.

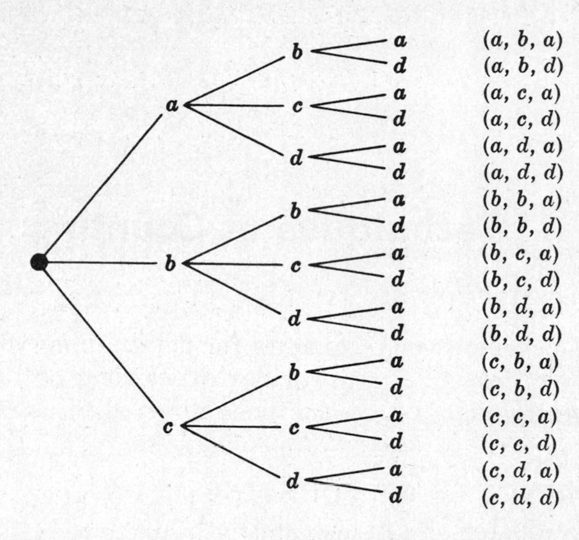

The elements of $S \times T \times W$ are the ordered triplets listed to the right of the tree diagram.

1.48. Each has 60 elements.

1.49. Both are true: $A \times A = (B \times B) \cap (C \times C) = (B \times C) \cap (C \times B)$.

1.51. (i) A_{14}, (ii) A_{24}, (iii) A_3, (iv) A_{12}, (v) A_s, (vi) A_{st}.

1.52. $\mathcal{P}(B) = \{B, \{1, \{2,3\}\}, \{1,4\}, \{\{2,3\},4\}, \{1\}, \{\{2,3\}\}, \{4\}, \emptyset\}$.

1.53. (i) no, (ii) no, (iii) yes, (iv) yes.

1.54. $[\{1,2,3\}]$, $[\{1\}, \{2,3\}]$, $[\{2\}, \{1,3\}]$, $[\{3\}, \{1,2\}]$ and $[\{1\}, \{2\}, \{3\}]$

Chapter 2

Techniques of Counting

INTRODUCTION

In this chapter we develop some techniques for determining without direct enumeration the number of possible outcomes of a particular experiment or the number of elements in a particular set. Such techniques are sometimes referred to as combinatorial analysis.

FUNDAMENTAL PRINCIPLE OF COUNTING

We begin with the following basic principle.

Fundamental Principle of Counting: If some procedure can be performed in n_1 different ways, and if, following this procedure, a second procedure can be performed in n_2 different ways, and if, following this second procedure, a third procedure can be performed in n_3 different ways, and so forth; then the number of ways the procedures can be performed in the order indicated is the product $n_1 \cdot n_2 \cdot n_3 \ldots$.

Example 2.1: Suppose a license plate contains two distinct letters followed by three digits with the first digit not zero. How many different license plates can be printed?

The first letter can be printed in 26 different ways, the second letter in 25 different ways (since the letter printed first cannot be chosen for the second letter), the first digit in 9 ways and each of the other two digits in 10 ways. Hence

$$26 \cdot 25 \cdot 9 \cdot 10 \cdot 10 = 585{,}000$$

different plates can be printed.

FACTORIAL NOTATION

The product of the positive integers from 1 to n inclusive occurs very often in mathematics and hence is denoted by the special symbol $n!$ (read "n factorial"):

$$n! = 1 \cdot 2 \cdot 3 \cdot \cdots \cdot (n-2)(n-1)n$$

It is also convenient to define $0! = 1$.

Example 2.2: $2! = 1 \cdot 2 = 2, \quad 3! = 1 \cdot 2 \cdot 3 = 6, \quad 4! = 1 \cdot 2 \cdot 3 \cdot 4 = 24,$
$5! = 5 \cdot 4! = 5 \cdot 24 = 120, \quad 6! = 6 \cdot 5! = 6 \cdot 120 = 720$

Example 2.3: $\frac{8!}{6!} = \frac{8 \cdot 7 \cdot 6!}{6!} = 8 \cdot 7 = 56 \qquad 12 \cdot 11 \cdot 10 = \frac{12 \cdot 11 \cdot 10 \cdot 9!}{9!} = \frac{12!}{9!}$

PERMUTATIONS

An arrangement of a set of n objects in a given order is called a *permutation* of the objects (taken all at a time). An arrangement of any $r \leq n$ of these objects in a given order is called an *r-permutation* or a *permutation of the n objects taken r at a time.*

Example 2.4: Consider the set of letters a, b, c and d. Then:

(i) *bdca, dcba* and *acdb* are permutations of the 4 letters (taken all at a time);

(ii) *bad, adb, cbd* and *bca* are permutations of the 4 letters taken 3 at a time;

(iii) *ad, cb, da* and *bd* are permutations of the 4 letters taken 2 at a time.

The number of permutations of n objects taken r at a time will be denoted by

$$P(n, r)$$

Before we derive the general formula for $P(n, r)$ we consider a special case.

Example 2.5: Find the number of permutations of 6 objects, say a, b, c, d, e, f, taken three at a time. In other words, find the number of "three letter words" with distinct letters that can be formed from the above six letters.

Let the general three letter word be represented by three boxes:

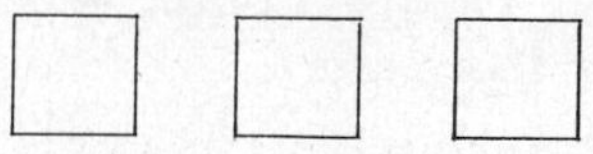

Now the first letter can be chosen in 6 different ways; following this, the second letter can be chosen in 5 different ways; and, following this, the last letter can be chosen in 4 different ways. Write each number in its appropriate box as follows:

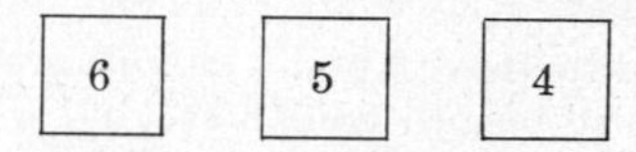

Thus by the fundamental principle of counting there are $6 \cdot 5 \cdot 4 = 120$ possible three letter words without repetitions from the six letters, or there are 120 permutations of 6 objects taken 3 at a time. That is,

$$P(6, 3) = 120$$

The derivation of the formula for $P(n, r)$ follows the procedure in the preceding example. The first element in an r-permutation of n-objects can be chosen in n different ways; following this, the second element in the permutation can be chosen in $n-1$ ways; and, following this, the third element in the permutation can be chosen in $n-2$ ways. Continuing in this manner, we have that the rth (last) element in the r-permutation can be chosen in $n-(r-1) = n-r+1$ ways. Thus

Theorem 2.1: $P(n, r) = n(n-1)(n-2)\cdots(n-r+1) = \dfrac{n!}{(n-r)!}$

The second part of the formula follows from the fact that

$$n(n-1)(n-2)\cdots(n-r+1) = \frac{n(n-1)(n-2)\cdots(n-r+1)\cdot(n-r)!}{(n-r)!} = \frac{n!}{(n-r)!}$$

In the special case that $r = n$, we have

$$P(n, n) = n(n-1)(n-2)\cdots 3\cdot 2\cdot 1 = n!$$

Namely,

Corollary 2.2: There are $n!$ permutations of n objects (taken all at a time).

Example 2.6: How many permutations are there of 3 objects, say, a, b and c?

By the above corollary there are $3! = 1\cdot 2\cdot 3 = 6$ such permutations. These are *abc, acb, bac, bca, cab, cba*.

PERMUTATIONS WITH REPETITIONS

Frequently we want to know the number of permutations of objects some of which are alike, as illustrated below. The general formula follows.

Theorem 2.3: The number of permutations of n objects of which n_1 are alike, n_2 are alike, ..., n_r are alike is

$$\frac{n!}{n_1!\, n_2! \cdots n_r!}$$

We indicate the proof of the above theorem by a particular example. Suppose we want to form all possible 5 letter words using the letters from the word DADDY. Now there are $5! = 120$ permutations of the objects D_1, A, D_2, D_3, Y where the three D's are distinguished. Observe that the following six permutations

$$D_1D_2D_3AY,\quad D_2D_1D_3AY,\quad D_3D_1D_2AY,\quad D_1D_3D_2AY,\quad D_2D_3D_1AY,\quad D_3D_2D_1AY$$

produce the same word when the subscripts are removed. The 6 comes from the fact that there are $3! = 3\cdot 2\cdot 1 = 6$ different ways of placing the three D's in the first three positions in the permutation. This is true for each of the other possible positions in which the D's appear. Accordingly there are

$$\frac{5!}{3!} = \frac{120}{6} = 20$$

different 5 letter words that can be formed using the letters from the word DADDY.

Example 2.7: How many different signals, each consisting of 8 flags hung in a vertical line, can be formed from a set of 4 indistinguishable red flags, 3 indistinguishable white flags, and a blue flag? We seek the number of permutations of 8 objects of which 4 are alike (the red flags) and 3 are alike (the white flags). By the above theorem, there are

$$\frac{8!}{4!\,3!} = \frac{8\cdot 7\cdot 6\cdot 5\cdot 4\cdot 3\cdot 2\cdot 1}{4\cdot 3\cdot 2\cdot 1\cdot 3\cdot 2\cdot 1} = 280$$

different signals.

ORDERED SAMPLES

Many problems in combinatorial analysis and, in particular, probability are concerned with choosing a ball from an urn containing n balls (or a card from a deck, or a person from a population). When we choose one ball after another from the urn, say r times, we call the choice an ordered sample of size r. We consider two cases:

(i) *Sampling with replacement.* Here the ball is replaced in the urn before the next ball is chosen. Now since there are n different ways to choose each ball, there are by the fundamental principle of counting

$$\overbrace{n\cdot n\cdot n\cdots n}^{r \text{ times}} = n^r$$

different ordered samples with replacement of size r.

(ii) *Sampling without replacement.* Here the ball is not replaced in the urn before the next ball is chosen. Thus there are no repetitions in the ordered sample. In other words, an ordered sample of size r without replacement is simply an r-permutation of the objects in the urn. Thus there are

$$P(n,r) = n(n-1)(n-2)\cdots(n-r+1) = \frac{n!}{(n-r)!}$$

different ordered samples of size r without replacement from a population of n objects.

Example 2.8: In how many ways can one choose three cards in succession from a deck of 52 cards (i) with replacement, (ii) without replacement? If each card is replaced in the deck before the next card is chosen, then each card can be chosen in 52 different ways. Hence there are

$$52\cdot 52\cdot 52 = 52^3 = 140{,}608$$

different ordered samples of size 3 with replacement.

On the other hand if there is no replacement, then the first card can be chosen in 52 different ways, the second card in 51 different ways, and the third and last card in 50 different ways. Thus there are

$$52 \cdot 51 \cdot 50 = 132{,}600$$

different ordered samples of size 3 without replacement.

BINOMIAL COEFFICIENTS AND THEOREM

The symbol $\binom{n}{r}$, read "nCr", where r and n are positive integers with $r \leq n$, is defined as follows:

$$\binom{n}{r} = \frac{n(n-1)(n-2)\cdots(n-r+1)}{1 \cdot 2 \cdot 3 \cdots (r-1)r}$$

These numbers are called the *binomial coefficients* in view of Theorem 2.5 below.

Example 2.9: $\binom{8}{2} = \frac{8 \cdot 7}{1 \cdot 2} = 28 \quad \binom{9}{4} = \frac{9 \cdot 8 \cdot 7 \cdot 6}{1 \cdot 2 \cdot 3 \cdot 4} = 126 \quad \binom{12}{5} = \frac{12 \cdot 11 \cdot 10 \cdot 9 \cdot 8}{1 \cdot 2 \cdot 3 \cdot 4 \cdot 5} = 792$

Observe that $\binom{n}{r}$ has exactly r factors in both the numerator and denominator. Also,

$$\binom{n}{r} = \frac{n(n-1)\cdots(n-r+1)}{1 \cdot 2 \cdot 3 \cdots (r-1)r} = \frac{n(n-1)\cdots(n-r+1)(n-r)!}{1 \cdot 2 \cdot 3 \cdots (r-1)r(n-r)!} = \frac{n!}{r!\,(n-r)!}$$

Using this formula and the fact that $n-(n-r) = r$, we obtain the following important relation.

Lemma 2.4: $\binom{n}{n-r} = \binom{n}{r}$ or, in other words, if $a+b=n$ then $\binom{n}{a} = \binom{n}{b}$.

Example 2.10: $\binom{10}{7} = \frac{10 \cdot 9 \cdot 8 \cdot 7 \cdot 6 \cdot 5 \cdot 4}{1 \cdot 2 \cdot 3 \cdot 4 \cdot 5 \cdot 6 \cdot 7} = 120$ or $\binom{10}{7} = \binom{10}{3} = \frac{10 \cdot 9 \cdot 8}{1 \cdot 2 \cdot 3} = 120$

Note that the second method saves both space and time.

Remark: Motivated by the second formula for $\binom{n}{r}$ and the fact that $0! = 1$, we define:

$$\binom{n}{0} = \frac{n!}{0!\,n!} = 1 \quad \text{and, in particular,} \quad \binom{0}{0} = \frac{0!}{0!\,0!} = 1$$

The Binomial Theorem, which is proved (Problem 2.18) by mathematical induction, gives the general expression for the expansion of $(a+b)^n$.

Theorem 2.5 (Binomial Theorem):

$$\begin{aligned}(a+b)^n &= \sum_{r=0}^{n} \binom{n}{r} a^{n-r}b^r \\ &= a^n + na^{n-1}b + \frac{n(n-1)}{1 \cdot 2}a^{n-2}b^2 + \cdots + nab^{n-1} + b^n\end{aligned}$$

Example 2.11:

$$\begin{aligned}(a+b)^5 &= a^5 + 5a^4b + \frac{5 \cdot 4}{1 \cdot 2}a^3b^2 + \frac{5 \cdot 4}{1 \cdot 2}a^2b^3 + 5ab^4 + b^5 \\ &= a^5 + 5a^4b + 10a^3b^2 + 10a^2b^3 + 5ab^4 + b^5\end{aligned}$$

$$\begin{aligned}(a+b)^6 &= a^6 + 6a^5b + \frac{6 \cdot 5}{1 \cdot 2}a^4b^2 + \frac{6 \cdot 5 \cdot 4}{1 \cdot 2 \cdot 3}a^3b^3 + \frac{6 \cdot 5}{1 \cdot 2}a^2b^4 + 6ab^5 + b^6 \\ &= a^6 + 6a^5b + 15a^4b^2 + 20a^3b^3 + 15a^2b^4 + 6ab^5 + b^6\end{aligned}$$

The following properties of the expansion of $(a+b)^n$ should be observed:

(i) There are $n+1$ terms.

(ii) The sum of the exponents of a and b in each term is n.

(iii) The exponents of a decrease term by term from n to 0; the exponents of b increase term by term from 0 to n.

(iv) The coefficient of any term is $\binom{n}{k}$ where k is the exponent of either a or b. (This follows from Lemma 2.4.)

(v) The coefficients of terms equidistant from the ends are equal.

We remark that the coefficients of the successive powers of $a+b$ can be arranged in a triangular array of numbers, called Pascal's triangle, as follows:

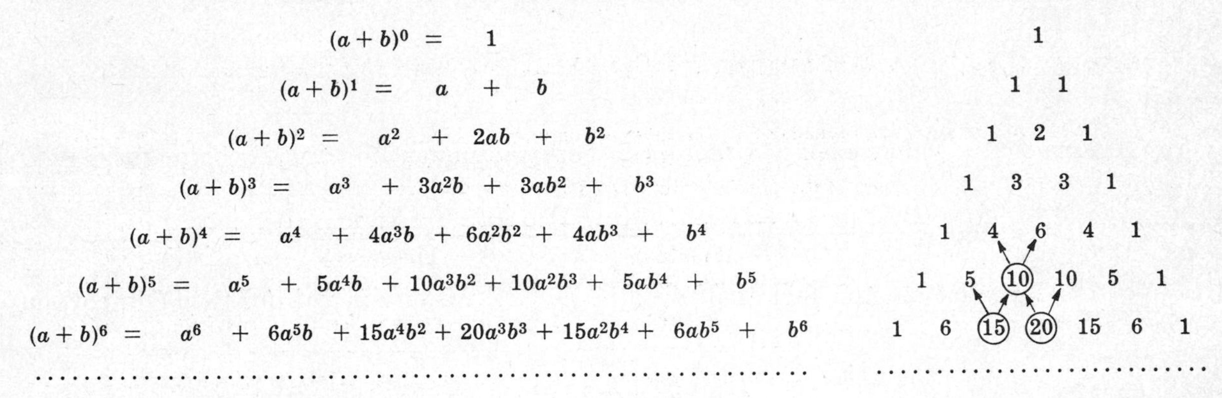

Pascal's triangle has the following interesting properties.

(*a*) The first number and the last number in each row is 1.

(*b*) Every other number in the array can be obtained by adding the two numbers appearing directly above it. For example: $10 = 4+6$, $15 = 5+10$, $20 = 10+10$.

We note that property (*b*) above is equivalent to the following theorem about binomial coefficients.

Theorem 2.6: $\binom{n+1}{r} = \binom{n}{r-1} + \binom{n}{r}$

Now let $n_1, n_2, \ldots, n_r$ be nonnegative integers such that $n_1 + n_2 + \cdots + n_r = n$. Then the expression $\binom{n}{n_1, n_2, \ldots, n_r}$ is defined as follows:

$$\binom{n}{n_1, n_2, \ldots, n_r} = \frac{n!}{n_1!\,n_2!\cdots n_r!}$$

For example,

$$\binom{7}{2, 3, 2} = \frac{7!}{2!3!2!} = 210 \qquad \binom{8}{4, 2, 2, 0} = \frac{8!}{4!2!2!0!} = 420$$

These numbers are called the *multinomial coefficients* in view of the following theorem which generalizes the binomial theorem.

Theorem 2.7: $(a_1 + a_2 + \cdots + a_r)^n = \sum_{n_1+n_2+\cdots+n_r=n} \binom{n}{n_1, n_2, \ldots, n_r} a_1^{n_1} a_2^{n_2} \cdots a_r^{n_r}$

COMBINATIONS

Suppose we have a collection of n objects. A *combination* of these n objects *taken* r *at a time*, or an r*-combination*, is any subset of r elements. In other words, an r-combination is any selection of r of the n objects where order does not count.

Example 2.12: The combinations of the letters a, b, c, d taken 3 at a time are

$$\{a, b, c\},\ \{a, b, d\},\ \{a, c, d\},\ \{b, c, d\} \quad \text{or simply} \quad abc,\ abd,\ acd,\ bcd$$

Observe that the following combinations are equal:

$$abc,\ acb,\ bac,\ bca,\ cab,\ cba$$

That is, each denotes the same set $\{a, b, c\}$.

The number of combinations of n objects taken r at a time will be denoted by

$$C(n, r)$$

Before we give the general formula for $C(n, r)$, we consider a special case.

Example 2.13: We determine the number of combinations of the four letters a, b, c, d taken 3 at a time. Note that each combination consisting of three letters determines $3! = 6$ permutations of the letters in the combination:

Combinations	Permutations
abc	*abc, acb, bac, bca, cab, cba*
abd	*abd, adb, bad, bda, dab, dba*
acd	*acd, adc, cad, cda, dac, dca*
bcd	*bcd, bdc, cbd, cdb, dbc, dcb*

Thus the number of combinations multiplied by $3!$ equals the number of permutations:

$$C(4, 3) \cdot 3! = P(4, 3) \qquad \text{or} \qquad C(4, 3) = \frac{P(4, 3)}{3!}$$

Now $P(4, 3) = 4 \cdot 3 \cdot 2 = 24$ and $3! = 6$; hence $C(4, 3) = 4$ as noted above.

Since each combination of n objects taken r at a time determines $r!$ permutations of the objects, we can conclude that

$$P(n, r) = r!\,C(n, r)$$

Thus we obtain

Theorem 2.8: $C(n, r) = \dfrac{P(n, r)}{r!} = \dfrac{n!}{r!\,(n-r)!}$

Recall that the binomial coefficient $\binom{n}{r}$ was defined to be $\dfrac{n!}{r!\,(n-r)!}$; hence

$$\boxed{C(n, r) = \binom{n}{r}}$$

We shall use $C(n, r)$ and $\binom{n}{r}$ interchangeably.

Example 2.14: How many committees of 3 can be formed from 8 people? Each committee is essentially a combination of the 8 people taken 3 at a time. Thus

$$C(8,3) = \binom{8}{3} = \frac{8\cdot 7\cdot 6}{1\cdot 2\cdot 3} = 56$$

different committees can be formed.

ORDERED PARTITIONS

Suppose an urn A contains seven marbles numbered 1 through 7. We compute the number of ways we can draw, first, 2 marbles from the urn, then 3 marbles from the urn, and lastly 2 marbles from the urn. In other words, we want to compute the number of *ordered partitions*

$$(A_1, A_2, A_3)$$

of the set of 7 marbles into cells A_1 containing 2 marbles, A_2 containing 3 marbles and A_3 containing 2 marbles. We call these ordered partitions since we distinguish between

$$(\{1,2\}\ \{3,4,5\},\ \{6,7\}) \quad\text{and}\quad (\{6,7\},\ \{3,4,5\},\ \{1,2\})$$

each of which yields the same partition of A.

Since we begin with 7 marbles in the urn, there are $\binom{7}{2}$ ways of drawing the first 2 marbles, i.e. of determining the first cell A_1; following this, there are 5 marbles left in the urn and so there are $\binom{5}{3}$ ways of drawing the 3 marbles, i.e. of determining A_2; finally, there are 2 marbles left in the urn and so there are $\binom{2}{2}$ ways of determining the last cell A_3. Thus there are

$$\binom{7}{2}\binom{5}{3}\binom{2}{2} = \frac{7\cdot 6}{1\cdot 2}\cdot\frac{5\cdot 4\cdot 3}{1\cdot 2\cdot 3}\cdot\frac{2\cdot 1}{1\cdot 2} = 210$$

different ordered partitions of A into cells A_1 containing 2 marbles, A_2 containing 3 marbles, and A_3 containing 2 marbles.

Now observe that

$$\binom{7}{2}\binom{5}{3}\binom{2}{2} = \frac{7!}{2!\,5!}\cdot\frac{5!}{3!\,2!}\cdot\frac{2!}{2!\,0!} = \frac{7!}{2!\,3!\,2!}$$

since each numerator after the first is cancelled by the second term in the denominator of the previous factor. In a similar manner we prove (Problem 2.28)

Theorem 2.9: Let A contain n elements and let $n_1, n_2, \ldots, n_r$ be positive integers with $n_1 + n_2 + \cdots + n_r = n$. Then there exist

$$\frac{n!}{n_1!\,n_2!\,n_3!\cdots n_r!}$$

different ordered partitions of A of the form $(A_1, A_2, \ldots, A_r)$ where A_1 contains n_1 elements, A_2 contains n_2 elements, ..., and A_r contains n_r elements.

Example 2.15: In how many ways can 9 toys be divided between 4 children if the youngest child is to receive 3 toys and each of the other children 2 toys?

We wish to find the number of ordered partitions of the 9 toys into 4 cells containing 3, 2, 2 and 2 toys respectively. By the above theorem, there are

$$\frac{9!}{3!\,2!\,2!\,2!} = 7560$$

such ordered partitions.

TREE DIAGRAMS

A tree diagram is a device used to enumerate all the possible outcomes of a sequence of experiments where each experiment can occur in a finite number of ways. The construction of tree diagrams is illustrated in the following examples.

Example 2.16: Find the product set $A \times B \times C$ where $A = \{1, 2\}$, $B = \{a, b, c\}$ and $C = \{3, 4\}$.

The tree diagram follows:

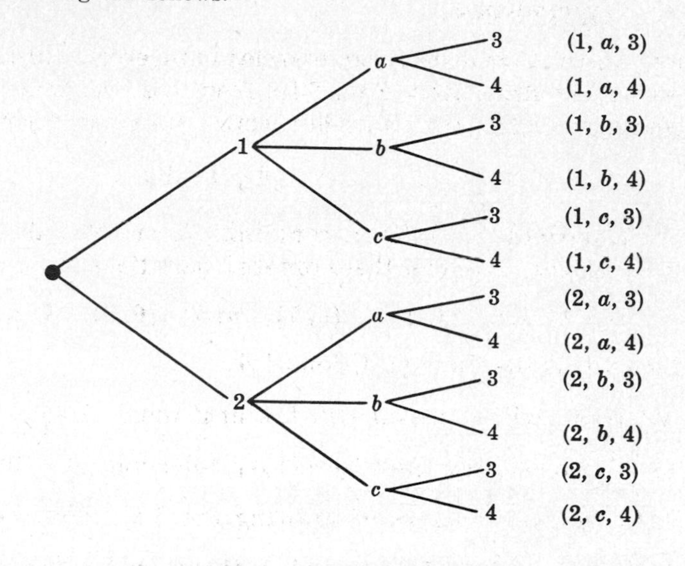

Observe that the tree is constructed from left to right, and that the number of branches at each point corresponds to the number of possible outcomes of the next experiment.

Example 2.17: Mark and Eric are to play a tennis tournament. The first person to win two games in a row or who wins a total of three games wins the tournament. The following diagram shows the possible outcomes of the tournament.

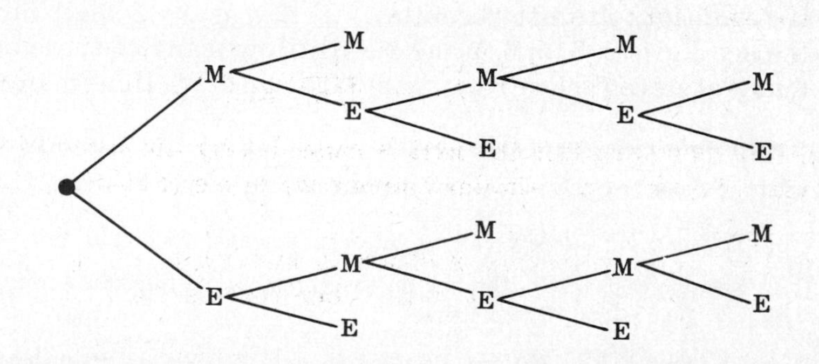

Observe that there are 10 endpoints which correspond to the 10 possible outcomes of the tournament:

MM, MEMM, MEMEM, MEMEE, MEE, EMM, EMEMM, EMEME, EMEE, EE

The path from the beginning of the tree to the endpoint indicates who won which game in the individual tournament.

Solved Problems

FACTORIAL

2.1. Compute $4!, 5!, 6!, 7!$ and $8!$.

$4! = 1\cdot2\cdot3\cdot4 = 24$

$5! = 1\cdot2\cdot3\cdot4\cdot5 = 5\cdot4! = 5\cdot24 = 120$

$6! = 1\cdot2\cdot3\cdot4\cdot5\cdot6 = 6\cdot5! = 6\cdot120 = 720$

$7! = 7\cdot6! = 7\cdot720 = 5040$

$8! = 8\cdot7! = 8\cdot5040 = 40{,}320$

2.2. Compute: (i) $\dfrac{13!}{11!}$, (ii) $\dfrac{7!}{10!}$.

(i) $$\frac{13!}{11!} = \frac{13\cdot12\cdot11\cdot10\cdot9\cdot8\cdot7\cdot6\cdot5\cdot4\cdot3\cdot2\cdot1}{11\cdot10\cdot9\cdot8\cdot7\cdot6\cdot5\cdot4\cdot3\cdot2\cdot1} = 13\cdot12 = 156$$

or $$\frac{13!}{11!} = \frac{13\cdot12\cdot11!}{11!} = 13\cdot12 = 156$$

(ii) $$\frac{7!}{10!} = \frac{7!}{10\cdot9\cdot8\cdot7!} = \frac{1}{10\cdot9\cdot8} = \frac{1}{720}$$

2.3. Simplify: (i) $\dfrac{n!}{(n-1)!}$, (ii) $\dfrac{(n+2)!}{n!}$.

(i) $$\frac{n!}{(n-1)!} = \frac{n(n-1)(n-2)\cdots3\cdot2\cdot1}{(n-1)(n-2)\cdots3\cdot2\cdot1} = n \quad \text{or, simply,} \quad \frac{n!}{(n-1)!} = \frac{n(n-1)!}{(n-1)!} = n$$

(ii) $$\frac{(n+2)!}{n!} = \frac{(n+2)(n+1)n(n-1)(n-2)\cdots3\cdot2\cdot1}{n(n-1)(n-2)\cdots3\cdot2\cdot1} = (n+2)(n+1) = n^2+3n+2$$

or, simply, $$\frac{(n+2)!}{n!} = \frac{(n+2)(n+1)\cdot n!}{n!} = (n+2)(n+1) = n^2+3n+2$$

PERMUTATIONS, ORDERED SAMPLES

2.4. If repetitions are not permitted, (i) how many 3 digit numbers can be formed from the six digits 2, 3, 5, 6, 7 and 9? (ii) How many of these are less than 400? (iii) How many are even? (iv) How many are odd? (v) How many are multiples of 5?

In each case draw three boxes [ ] [ ] [ ] to represent an arbitrary number, and then write in each box the number of digits that can be placed there.

(i) The box on the left can be filled in 6 ways; following this, the middle box can be filled in 5 ways; and, lastly, the box on the right can be filled in 4 ways: [6] [5] [4]. Thus there are $6\cdot5\cdot4 = 120$ numbers.

(ii) The box on the left can be filled in only 2 ways, by 2 or 3, since each number must be less than 400; the middle box can be filled in 5 ways; and, lastly, the box on the right can be filled in 4 ways: [2] [5] [4]. Thus there are $2\cdot5\cdot4 = 40$ numbers.

(iii) The box on the right can be filled in only 2 ways, by 2 or 6, since the numbers must be even; the box on the left can then be filled in 5 ways; and, lastly, the middle box can be filled in 4 ways: [5] [4] [2]. Thus there are $5\cdot4\cdot2 = 40$ numbers.

(iv) The box on the right can be filled in only 4 ways, by 3, 5, 7 or 9, since the numbers must be odd; the box on the left can then be filled in 5 ways; and, lastly, the box in the middle can be filled in 4 ways: [5] [4] [4]. Thus there are $5\cdot4\cdot4 = 80$ numbers.

(v) The box on the right can be filled in only 1 way, by 5, since the numbers must be multiples of 5; the box on the left can then be filled in 5 ways; and, lastly, the box in the middle can be filled in 4 ways: | 5 | 4 | 1 |. Thus there are $5 \cdot 4 \cdot 1 = 20$ numbers.

2.5. In how many ways can a party of 7 persons arrange themselves (i) in a row of 7 chairs? (ii) around a circular table?

(i) The seven persons can arrange themselves in a row in $7 \cdot 6 \cdot 5 \cdot 4 \cdot 3 \cdot 2 \cdot 1 = 7!$ ways.

(ii) One person can sit at any place in the circular table. The other six persons can then arrange themselves in $6 \cdot 5 \cdot 4 \cdot 3 \cdot 2 \cdot 1 = 6!$ ways around the table.

This is an example of a *circular permutation*. In general, n objects can be arranged in a circle in $(n-1)(n-2) \cdots 3 \cdot 2 \cdot 1 = (n-1)!$ ways.

2.6. (i) In how many ways can 3 boys and 2 girls sit in a row? (ii) In how many ways can they sit in a row if the boys and girls are each to sit together? (iii) In how many ways can they sit in a row if just the girls are to sit together?

(i) The five persons can sit in a row in $5 \cdot 4 \cdot 3 \cdot 2 \cdot 1 = 5! = 120$ ways.

(ii) There are 2 ways to distribute them according to sex: BBBGG or GGBBB. In each case the boys can sit in $3 \cdot 2 \cdot 1 = 3! = 6$ ways, and the girls can sit in $2 \cdot 1 = 2! = 2$ ways. Thus, altogether, there are $2 \cdot 3! \cdot 2! = 2 \cdot 6 \cdot 2 = 24$ ways.

(iii) There are 4 ways to distribute them according to sex: GGBBB, BGGBB, BBGGB, BBBGG. Note that each way corresponds to the number, 0, 1, 2 or 3, of boys sitting to the left of the girls. In each case, the boys can sit in $3!$ ways, and the girls in $2!$ ways. Thus, altogether, there are $4 \cdot 3! \cdot 2! = 4 \cdot 6 \cdot 2 = 48$ ways.

2.7. How many different signals, each consisting of 6 flags hung in a vertical line, can be formed from 4 identical red flags and 2 identical blue flags?

This problem concerns permutations with repetitions. There are $\frac{6!}{4!\,2!} = 15$ signals since there are 6 flags of which 4 are red and 2 are blue.

2.8. How many distinct permutations can be formed from all the letters of each word: (i) them, (ii) unusual, (iii) sociological?

(i) $4! = 24$, since there are 4 letters and no repetitions.

(ii) $\frac{7!}{3!} = 840$, since there are 7 letters of which 3 are u.

(iii) $\frac{12!}{3!\,2!\,2!\,2!}$, since there are 12 letters of which 3 are o, 2 are c, 2 are i, and 2 are l.

2.9. (i) In how many ways can 3 Americans, 4 Frenchmen, 4 Danes and 2 Italians be seated in a row so that those of the same nationality sit together?

(ii) Solve the same problem if they sit at a round table.

(i) The 4 nationalities can be arranged in a row in $4!$ ways. In each case the 3 Americans can be seated in $3!$ ways, the 4 Frenchmen in $4!$ ways, the 4 Danes in $4!$ ways, and the 2 Italians in $2!$ ways. Thus, altogether, there are $4!\,3!\,4!\,4!\,2! = 165{,}888$ arrangements.

(ii) The 4 nationalities can be arranged in a circle in $3!$ ways (see Problem 14.4 on circular permutations). In each case the 3 Americans can be seated in $3!$ ways, the 4 Frenchmen in $4!$ ways, the 4 Danes in $4!$ ways, and the 2 Italians in $2!$ ways. Thus, altogether, there are $3!\,3!\,4!\,4!\,2! = 41{,}472$ arrangements.

2.10. Suppose an urn contains 8 balls. Find the number of ordered samples of size 3 (i) with replacement, (ii) without replacement.

(i) Each ball in the ordered sample can be chosen in 8 ways; hence there are $8 \cdot 8 \cdot 8 = 8^3 = 512$ samples with replacement.

(ii) The first ball in the ordered sample can be chosen in 8 ways, the next in 7 ways, and the last in 6 ways. Thus there are $8 \cdot 7 \cdot 6 = 336$ samples without replacement.

2.11. Find n if (i) $P(n,2) = 72$, (ii) $P(n,4) = 42P(n,2)$, (iii) $2P(n,2) + 50 = P(2n,2)$.

(i) $P(n,2) = n(n-1) = n^2 - n$; hence $n^2 - n = 72$ or $n^2 - n - 72 = 0$ or $(n-9)(n+8) = 0$.

Since n must be positive, the only answer is $n = 9$.

(ii) $P(n,4) = n(n-1)(n-2)(n-3)$ and $P(n,2) = n(n-1)$. Hence

$$n(n-1)(n-2)(n-3) = 42n(n-1) \quad \text{or, if } n \neq 0, \neq 1, \quad (n-2)(n-3) = 42$$

$$\text{or} \quad n^2 - 5n + 6 = 42 \quad \text{or} \quad n^2 - 5n - 36 = 0 \quad \text{or} \quad (n-9)(n+4) = 0$$

Since n must be positive, the only answer is $n = 9$.

(iii) $P(n,2) = n(n-1) = n^2 - n$ and $P(2n,2) = 2n(2n-1) = 4n^2 - 2n$. Hence

$$2(n^2 - n) + 50 = 4n^2 - 2n \quad \text{or} \quad 2n^2 - 2n + 50 = 4n^2 - 2n \quad \text{or} \quad 50 = 2n^2 \quad \text{or} \quad n^2 = 25$$

Since n must be positive, the only answer is $n = 5$.

BINOMIAL COEFFICIENTS AND THEOREM

2.12. Compute: (i) $\binom{16}{3}$, (ii) $\binom{12}{4}$, (iii) $\binom{15}{5}$.

Recall that there are as many factors in the numerator as in the denominator.

(i) $\binom{16}{3} = \dfrac{16 \cdot 15 \cdot 14}{1 \cdot 2 \cdot 3} = 560$

(iii) $\binom{15}{5} = \dfrac{15 \cdot 14 \cdot 13 \cdot 12 \cdot 11}{1 \cdot 2 \cdot 3 \cdot 4 \cdot 5} = 3003$

(ii) $\binom{12}{4} = \dfrac{12 \cdot 11 \cdot 10 \cdot 9}{1 \cdot 2 \cdot 3 \cdot 4} = 495$

2.13. Compute: (i) $\binom{8}{5}$, (ii) $\binom{9}{7}$, (iii) $\binom{10}{6}$.

(i) $\binom{8}{5} = \dfrac{8 \cdot 7 \cdot 6 \cdot 5 \cdot 4}{1 \cdot 2 \cdot 3 \cdot 4 \cdot 5} = 56$

Note that $8 - 5 = 3$; hence we could also compute $\binom{8}{5}$ as follows:

$$\binom{8}{5} = \binom{8}{3} = \frac{8 \cdot 7 \cdot 6}{1 \cdot 2 \cdot 3} = 56$$

(ii) Now $9 - 7 = 2$; hence $\binom{9}{7} = \binom{9}{2} = \dfrac{9 \cdot 8}{1 \cdot 2} = 36.$

(iii) Now $10 - 6 = 4$; hence $\binom{10}{6} = \binom{10}{4} = \dfrac{10 \cdot 9 \cdot 8 \cdot 7}{1 \cdot 2 \cdot 3 \cdot 4} = 210.$

2.14. Expand and simplify: $(2x + y^2)^5$.

$$\begin{aligned}(2x+y^2)^5 &= (2x)^5 + \frac{5}{1}(2x)^4(y^2) + \frac{5 \cdot 4}{1 \cdot 2}(2x)^3(y^2)^2 + \frac{5 \cdot 4}{1 \cdot 2}(2x)^2(y^2)^3 + \frac{5}{1}(2x)(y^2)^4 + (y^2)^5 \\ &= 32x^5 + 80x^4y^2 + 80x^3y^4 + 40x^2y^6 + 10xy^8 + y^{10}\end{aligned}$$

2.15. Expand and simplify: $(x^2-2y)^6$.

$$\begin{aligned}(x^2-2y)^6 &= (x^2)^6 + \frac{6}{1}(x^2)^5(-2y) + \frac{6\cdot 5}{1\cdot 2}(x^2)^4(-2y)^2 + \frac{6\cdot 5\cdot 4}{1\cdot 2\cdot 3}(x^2)^3(-2y)^3 \\ &\quad + \frac{6\cdot 5}{1\cdot 2}(x^2)^2(-2y)^4 + \frac{6}{1}(x^2)(-2y)^5 + (-2y)^6 \\ &= x^{12} - 12x^{10}y + 60x^8y^2 - 160x^6y^3 + 240x^4y^4 - 192x^2y^5 + 64y^6\end{aligned}$$

2.16. Prove: $2^4 = 16 = \binom{4}{0} + \binom{4}{1} + \binom{4}{2} + \binom{4}{3} + \binom{4}{4}$.

Expand $(1+1)^4$ using the binomial theorem:

$$\begin{aligned}2^4 = (1+1)^4 &= \binom{4}{0}1^4 + \binom{4}{1}1^3\,1^1 + \binom{4}{2}1^2\,1^2 + \binom{4}{3}1^1\,1^3 + \binom{4}{4}1^4 \\ &= \binom{4}{0} + \binom{4}{1} + \binom{4}{2} + \binom{4}{3} + \binom{4}{4}\end{aligned}$$

2.17. Prove Theorem 2.6: $\binom{n+1}{r} = \binom{n}{r-1} + \binom{n}{r}$.

Now $\binom{n}{r-1} + \binom{n}{r} = \frac{n!}{(r-1)!\cdot(n-r+1)!} + \frac{n!}{r!\cdot(n-r)!}$. To obtain the same denominator in both fractions, multiply the first fraction by $\frac{r}{r}$ and the second fraction by $\frac{n-r+1}{n-r+1}$. Hence

$$\begin{aligned}\binom{n}{r-1} + \binom{n}{r} &= \frac{r\cdot n!}{r\cdot(r-1)!\cdot(n-r+1)!} + \frac{(n-r+1)\cdot n!}{r!\cdot(n-r+1)\cdot(n-r)!} \\ &= \frac{r\cdot n!}{r!\,(n-r+1)!} + \frac{(n-r+1)\cdot n!}{r!\,(n-r+1)!} \\ &= \frac{r\cdot n! + (n-r+1)\cdot n!}{r!\,(n-r+1)!} = \frac{[r+(n-r+1)]\cdot n!}{r!\,(n-r+1)!} \\ &= \frac{(n+1)n!}{r!\,(n-r+1)!} = \frac{(n+1)!}{r!\,(n-r+1)!} = \binom{n+1}{r}\end{aligned}$$

2.18. Prove the Binomial Theorem 2.5: $(a+b)^n = \sum_{r=0}^{n}\binom{n}{r}a^{n-r}b^r$.

The theorem is true for $n=1$, since

$$\sum_{r=0}^{1}\binom{1}{r}a^{1-r}b^r = \binom{1}{0}a^1b^0 + \binom{1}{1}a^0b^1 = a + b = (a+b)^1$$

We assume the theorem holds for $(a+b)^n$ and prove it is true for $(a+b)^{n+1}$.

$$\begin{aligned}(a+b)^{n+1} &= (a+b)(a+b)^n \\ &= (a+b)\left[a^n + \binom{n}{1}a^{n-1}b + \cdots + \binom{n}{r-1}a^{n-r+1}b^{r-1}\right. \\ &\quad \left. + \binom{n}{r}a^{n-r}b^r + \cdots + \binom{n}{1}ab^{n-1} + b^n\right]\end{aligned}$$

Now the term in the product which contains b^r is obtained from

$$\begin{aligned}b\left[\binom{n}{r-1}a^{n-r+1}b^{r-1}\right] + a\left[\binom{n}{r}a^{n-r}b^r\right] &= \binom{n}{r-1}a^{n-r+1}b^r + \binom{n}{r}a^{n-r+1}b^r \\ &= \left[\binom{n}{r-1} + \binom{n}{r}\right]a^{n-r+1}b^r\end{aligned}$$

But, by Theorem 2.6, $\binom{n}{r-1} + \binom{n}{r} = \binom{n+1}{r}$. Thus the term containing b^r is $\binom{n+1}{r} a^{n-r+1} b^r$. Note that $(a+b)(a+b)^n$ is a polynomial of degree $n+1$ in b. Consequently,

$$(a+b)^{n+1} \quad = \quad (a+b)(a+b)^n \quad = \quad \sum_{r=0}^{n+1} \binom{n+1}{r} a^{n-r+1} b^r$$

which was to be proved.

2.19. Compute the following multinomial coefficients:

$$\text{(i)} \binom{6}{3,2,1}, \quad \text{(ii)} \binom{8}{4,2,2,0}, \quad \text{(iii)} \binom{10}{5,3,2,2}$$

(i) $\binom{6}{3,2,1} = \frac{6!}{3!\,2!\,1!} = \frac{6 \cdot 5 \cdot 4 \cdot 3 \cdot 2 \cdot 1}{3 \cdot 2 \cdot 1 \cdot 2 \cdot 1 \cdot 1} = 60$

(ii) $\binom{8}{4,2,2,0} = \frac{8!}{4!\,2!\,2!\,0!} = \frac{8 \cdot 7 \cdot 6 \cdot 5 \cdot 4 \cdot 3 \cdot 2 \cdot 1}{4 \cdot 3 \cdot 2 \cdot 1 \cdot 2 \cdot 1 \cdot 2 \cdot 1 \cdot 1} = 420$

(iii) The expression $\binom{10}{5,3,2,2}$ has no meaning since $5+3+2+2 \neq 10$.

COMBINATIONS

2.20. In how many ways can a committee consisting of 3 men and 2 women be chosen from 7 men and 5 women?

The 3 men can be chosen from the 7 men in $\binom{7}{3}$ ways, and the 2 women can be chosen from the 5 women in $\binom{5}{2}$ ways. Hence the committee can be chosen in $\binom{7}{3}\binom{5}{2} = \frac{7 \cdot 6 \cdot 5}{1 \cdot 2 \cdot 3} \cdot \frac{5 \cdot 4}{1 \cdot 2} = 350$ ways.

2.21. A delegation of 4 students is selected each year from a college to attend the National Student Association annual meeting. (i) In how many ways can the delegation be chosen if there are 12 eligible students? (ii) In how many ways if two of the eligible students will not attend the meeting together? (iii) In how many ways if two of the eligible students are married and will only attend the meeting together?

(i) The 4 students can be chosen from the 12 students in $\binom{12}{4} = \frac{12 \cdot 11 \cdot 10 \cdot 9}{1 \cdot 2 \cdot 3 \cdot 4} = 495$ ways.

(ii) Let A and B denote the students who will not attend the meeting together.

Method 1.

If neither A nor B is included, then the delegation can be chosen in $\binom{10}{4} = \frac{10 \cdot 9 \cdot 8 \cdot 7}{1 \cdot 2 \cdot 3 \cdot 4} =$ 210 ways. If either A or B, but not both, is included, then the delegation can be chosen in $2 \cdot \binom{10}{3} = 2 \cdot \frac{10 \cdot 9 \cdot 8}{1 \cdot 2 \cdot 3} = 240$ ways. Thus, altogether, the delegation can be chosen in $210 + 240 = 450$ ways.

Method 2.

If A and B are both included, then the other 2 members of the delegation can be chosen in $\binom{10}{2} = 45$ ways. Thus there are $495 - 45 = 450$ ways the delegation can be chosen if A and B are not both included.

(iii) Let C and D denote the married students. If C and D do not go, then the delegation can be chosen in $\binom{10}{4} = 210$ ways. If both C and D go, then the delegation can be chosen in $\binom{10}{2} = 45$ ways. Altogether, the delegation can be chosen in $210 + 45 = 255$ ways.

2.22. A student is to answer 8 out of 10 questions on an exam. (i) How many choices has he? (ii) How many if he must answer the first 3 questions? (iii) How many if he must answer at least 4 of the first 5 questions?

(i) The 8 questions can be selected in $\binom{10}{8} = \binom{10}{2} = \frac{10 \cdot 9}{1 \cdot 2} = 45$ ways.

(ii) If he answers the first 3 questions, then he can choose the other 5 questions from the last 7 questions in $\binom{7}{5} = \binom{7}{2} = \frac{7 \cdot 6}{1 \cdot 2} = 21$ ways.

(iii) If he answers all the first 5 questions, then he can choose the other 3 questions from the last 5 in $\binom{5}{3} = 10$ ways. On the other hand, if he answers only 4 of the first 5 questions, then he can choose these 4 in $\binom{5}{4} = \binom{5}{1} = 5$ ways, and he can choose the other 4 questions from the last 5 in $\binom{5}{4} = \binom{5}{1} = 5$ ways; hence he can choose the 8 questions in $5 \cdot 5 = 25$ ways. Thus he has a total of 35 choices.

2.23. Find the number of subsets of a set X containing n elements.

Method 1.

The number of subsets of X with $r \leq n$ elements is given by $\binom{n}{r}$. Hence, altogether, there are

$$\binom{n}{0} + \binom{n}{1} + \binom{n}{2} + \cdots + \binom{n}{n-1} + \binom{n}{n}$$

subsets of X. The above sum (Problem 2.51) is equal to 2^n, i.e. there are 2^n subsets of X.

Method 2.

There are two possibilities for each element of X: either it belongs to the subset or it doesn't; hence there are

$$\overbrace{2 \cdot 2 \cdot \cdots \cdot 2}^{n \text{ times}} = 2^n$$

ways to form a subset of X, i.e. there are 2^n different subsets of X.

2.24. In how many ways can a teacher choose one or more students from six eligible students?

Method 1.

By the preceding problem, there are $2^6 = 64$ subsets of the set consisting of the six students. However, the empty set must be deleted since one or more students are chosen. Accordingly there are $2^6 - 1 = 64 - 1 = 63$ ways to choose the students.

Method 2.

Either 1, 2, 3, 4, 5 or 6 students are chosen. Hence the number of choices is

$$\binom{6}{1} + \binom{6}{2} + \binom{6}{3} + \binom{6}{4} + \binom{6}{5} + \binom{6}{6} = 6 + 15 + 20 + 15 + 6 + 1 = 63$$

ORDERED AND UNORDERED PARTITIONS

2.25. In how many ways can 7 toys be divided among 3 children if the youngest gets 3 toys and each of the others gets 2?

We seek the number of ordered partitions of 7 objects into cells containing 3, 2 and 2 objects, respectively. By Theorem 2.9, there are $\frac{7!}{3!\,2!\,2!} = 210$ such partitions.

2.26. There are 12 students in a class. In how many ways can the 12 students take 3 different tests if 4 students are to take each test?

Method 1.

We seek the number of ordered partitions of the 12 students into cells containing 4 students each. By Theorem 2.9, there are $\frac{12!}{4!\,4!\,4!} = 34{,}650$ such partitions.

Method 2.

There are $\binom{12}{4}$ ways to choose 4 students to take the first test; following this, there are $\binom{8}{4}$ ways to choose 4 students to take the second test. The remaining students take the third test. Thus, altogether, there are $\binom{12}{4} \cdot \binom{8}{4} = 495 \cdot 70 = 34{,}650$ ways for the students to take the tests.

2.27. In how many ways can 12 students be partitioned into 3 teams, A_1, A_2 and A_3, so that each team contains 4 students?

Method 1.

Observe that each partition $\{A_1, A_2, A_3\}$ of the students can be arranged in $3! = 6$ ways as an ordered partition. Since (see the preceding problem) there are $\frac{12!}{4!\,4!\,4!} = 34{,}650$ such ordered partitions, there are $34{,}650/6 = 5775$ (unordered) partitions.

Method 2.

Let A denote one of the students. Then there are $\binom{11}{3}$ ways to choose 3 other students to be on the same team as A. Now let B denote a student who is not on the same team as A; then there are $\binom{7}{3}$ ways to choose 3 students of the remaining students to be on the same team as B. The remaining 4 students constitute the third team. Thus, altogether, there are $\binom{11}{3} \cdot \binom{7}{3} = 165 \cdot 35 = 5775$ ways to partition the students.

2.28. Prove Theorem 2.9: Let A contain n elements and let $n_1, n_2, \ldots, n_r$ be positive integers with $n_1 + n_2 + \cdots + n_r = n$. Then there exist

$$\frac{n!}{n_1!\, n_2!\, n_3! \cdots n_r!}$$

different ordered partitions of A of the form $(A_1, A_2, \ldots, A_r)$ where A_1 contains n_1 elements, A_2 contains n_2 elements, ..., and A_r contains n_r elements.

We begin with n elements in A; hence there are $\binom{n}{n_1}$ ways of selecting the cell A_1. Following this, there are $n - n_1$ elements left, i.e. in $A \setminus A_1$, and so there are $\binom{n-n_1}{n_2}$ ways of selecting A_2. Similarly, for $i = 3, \ldots, r$, there are $\binom{n - n_1 - \cdots - n_{i-1}}{n_i}$ ways of selecting A_i. Thus there are

$$\binom{n}{n_1}\binom{n-n_1}{n_2}\binom{n-n_1-n_2}{n_3} \cdots \binom{n-n_1-\cdots-n_{r-1}}{n_r} \qquad (*)$$

different ordered partitions of A. Now (*) is equal to

$$\frac{n!}{n_1!\,(n-n_1)!} \cdot \frac{(n-n_1)!}{n_2!\,(n-n_1-n_2)!} \cdot \cdots \cdot \frac{(n-n_1-\cdots-n_{r-1})!}{n_r!\,(n-n_1-\cdots-n_r)!}$$

But this is equal to $\frac{n!}{n_1!\,n_2! \cdots n_r!}$ since each numerator after the first is cancelled by the second term in the denominator and since $(n - n_1 - \cdots - n_r)! = 0! = 1$. Thus the theorem is proved.

TREE DIAGRAMS

2.29. Construct the tree diagram for the number of permutations of $\{a, b, c\}$.

a	b	c	abc
	c	b	acb
b	a	c	bac
	c	a	bca
c	a	b	cab
	b	a	cba

The six permutations are listed on the right of the diagram.

2.30. A man has time to play roulette at most five times. At each play he wins or loses a dollar. The man begins with one dollar and will stop playing before the five times if he loses all his money or if he wins three dollars, i.e. if he has four dollars. Find the number of ways that the betting can occur.

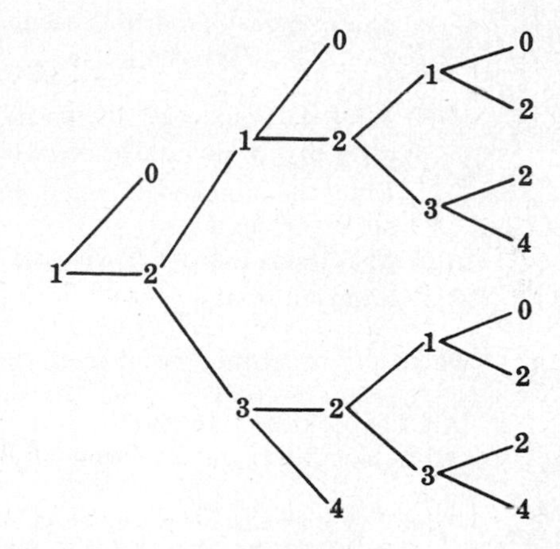

The tree diagram on the right describes the way the betting can occur. Each number in the diagram denotes the number of dollars the man has at that point. Observe that the betting can occur in 11 different ways. Note that he will stop betting before the five times are up in only three of the cases.

Supplementary Problems

FACTORIAL

2.31. Compute: (i) $9!$, (ii) $10!$, (iii) $11!$

2.32. Compute: (i) $\frac{16!}{14!}$, (ii) $\frac{14!}{11!}$, (iii) $\frac{8!}{10!}$, (iv) $\frac{10!}{13!}$.

2.33. Simplify: (i) $\frac{(n+1)!}{n!}$, (ii) $\frac{n!}{(n-2)!}$, (iii) $\frac{(n-1)!}{(n+2)!}$, (iv) $\frac{(n-r+1)!}{(n-r-1)!}$.

PERMUTATIONS

2.34. (i) How many automobile license plates can be made if each plate contains 2 different letters followed by 3 different digits? (ii) Solve the problem if the first digit cannot be 0.

2.35. There are 6 roads between A and B and 4 roads between B and C.

(i) In how many ways can one drive from A to C by way of B?

(ii) In how many ways can one drive roundtrip from A to C by way of B?

(iii) In how many ways can one drive roundtrip from A to C without using the same road more than once?

2.36. Find the number of ways in which 6 people can ride a toboggan if one of three must drive.

2.37. (i) Find the number of ways in which five persons can sit in a row.
(ii) How many ways are there if two of the persons insist on sitting next to one another?

2.38. Solve the preceding problem if they sit around a circular table.

2.39. (i) Find the number of four letter words that can be formed from the letters of the word HISTORY. (ii) How many of them contain only consonants? (iii) How many of them begin and end in a consonant? (iv) How many of them begin with a vowel? (v) How many contain the letter Y? (vi) How many begin with T and end in a vowel? (vii) How many begin with T and also contain S? (viii) How many contain both vowels?

2.40. How many different signals, each consisting of 8 flags hung in a vertical line, can be formed from 4 red flags, 2 blue flags and 2 green flags?

2.41. Find the number of permutations that can be formed from all the letters of each word: (i) queue, (ii) committee, (iii) proposition, (iv) baseball.

2.42. (i) Find the number of ways in which 4 boys and 4 girls can be seated in a row if the boys and girls are to have alternate seats.
(ii) Find the number of ways if they sit alternately and if one boy and one girl are to sit in adjacent seats.
(iii) Find the number of ways if they sit alternately and if one boy and one girl must not sit in adjacent seats.

2.43. Solve the preceding problem if they sit around a circular table.

2.44. An urn contains 10 balls. Find the number of ordered samples (i) of size 3 with replacement, (ii) of size 3 without replacement, (iii) of size 4 with replacement, (iv) of size 5 without replacement.

2.45. Find the number of ways in which 5 large books, 4 medium-size books and 3 small books can be placed on a shelf so that all books of the same size are together.

2.46. Consider all positive integers with 3 different digits. (Note that 0 cannot be the first digit.) (i) How many are greater than 700? (ii) How many are odd? (iii) How many are even? (iv) How many are divisible by 5?

2.47. (i) Find the number of distinct permutations that can be formed from all of the letters of the word ELEVEN. (ii) How many of them begin and end with E? (iii) How many of them have the 3 E's together? (iv) How many begin with E and end with N?

BINOMIAL COEFFICIENTS AND THEOREM

2.48. Compute: (i) $\binom{5}{2}$, (ii) $\binom{7}{3}$, (iii) $\binom{14}{2}$, (iv) $\binom{6}{4}$, (v) $\binom{20}{17}$, (vi) $\binom{18}{15}$.

2.49. Compute: (i) $\binom{9}{3,\,5,\,1}$, (ii) $\binom{7}{3,\,2,\,2,\,0}$, (iii) $\binom{6}{2,\,2,\,1,\,1,\,0}$.

2.50. Expand and simplify: (i) $(2x+y^2)^3$, (ii) $(x^2-3y)^4$, (iii) $(\frac{1}{2}a+2b)^5$, (iv) $(2a^2-b)^6$.

2.51. Show that $\binom{n}{0}+\binom{n}{1}+\binom{n}{2}+\binom{n}{3}+\cdots+\binom{n}{n} = 2^n.$

2.52. Show that $\binom{n}{0}-\binom{n}{1}+\binom{n}{2}-\binom{n}{3}+\cdots\pm\binom{n}{n} = 0.$

2.53. Find the term in the expansion of $(2x^2-\frac{1}{2}y^3)^8$ which contains x^8.

2.54. Find the term in the expansion of $(3xy^2-z^2)^7$ which contains y^6.

COMBINATIONS

2.55. A class contains 9 boys and 3 girls. (i) In how many ways can the teacher choose a committee of 4? (ii) How many of them will contain at least one girl? (iii) How many of them will contain exactly one girl?

2.56. A woman has 11 close friends. (i) In how many ways can she invite 5 of them to dinner? (ii) In how many ways if two of the friends are married and will not attend separately? (iii) In how many ways if two of them are not on speaking terms and will not attend together?

2.57. There are 10 points $A, B, \ldots$ in a plane, no three on the same line. (i) How many lines are determined by the points? (ii) How many of these lines do not pass through A or B? (iii) How many triangles are determined by the points? (iv) How many of these triangles contain the point A? (v) How many of these triangles contain the side AB?

2.58. A student is to answer 10 out of 13 questions on an exam. (i) How many choices has he? (ii) How many if he must answer the first two questions? (iii) How many if he must answer the first or second question but not both? (iv) How many if he must answer exactly 3 of the first 5 questions? (v) How many if he must answer at least 3 of the first 5 questions?

2.59. A man is dealt a poker hand (5 cards) from an ordinary playing deck. In how many ways can he be dealt (i) a straight flush, (ii) four of a kind, (iii) a straight, (iv) a pair of aces, (v) two of a kind (a pair)?

2.60. The English alphabet has 26 letters of which 5 are vowels.

(i) How many 5 letter words containing 3 different consonants and 2 different vowels can be formed?

(ii) How many of them contain the letter b?

(iii) How many of them contain the letters b and c?

(iv) How many of them begin with b and contain the letter c?

(v) How many of them begin with b and end with c?

(vi) How many of them contain the letters a and b?

(vii) How many of them begin with a and contain b?

(viii) How many of them begin with b and contain a?

(ix) How many of them begin with a and end with b?

(x) How many of them contain the letters a, b and c?

ORDERED AND UNORDERED PARTITIONS

2.61. In how many ways can 9 toys be divided evenly among 3 children?

2.62. In how many ways can 9 students be evenly divided into three teams?

2.63. In how many ways can 10 students be divided into three teams, one containing 4 students and the others 3?

2.64. There are 12 balls in an urn. In how many ways can 3 balls be drawn from the urn, four times in succession, all without replacement?

2.65. In how many ways can a club with 12 members be partitioned into three committees containing 5, 4 and 3 members respectively?

2.66. In how many ways can n students be partitioned into two teams containing at least one student?

2.67. In how many ways can 14 men be partitioned into 6 committees where 2 of the committees contain 3 men and the others 2?

TREE DIAGRAMS

2.68. Construct the tree diagram for the number of permutations of $\{a, b, c, d\}$.

2.69. Find the product set $\{1, 2, 3\} \times \{2, 4\} \times \{2, 3, 4\}$ by constructing the appropriate tree diagram.

2.70. Teams A and B play in a basketball tournament. The first team that wins two games in a row or a total of four games wins the tournament. Find the number of ways the tournament can occur.

2.71. A man has time to play roulette five times. He wins or loses a dollar at each play. The man begins with two dollars and will stop playing before the five times if he loses all his money or wins three dollars (i.e. has five dollars). Find the number of ways the playing can occur.

2.72. A man is at the origin on the x-axis and takes a unit step either to the left or to the right. He stops after 5 steps or if he reaches 3 or -2. Construct the tree diagram to describe all possible paths the man can travel.

2.73. In the following diagram let $A, B, \ldots, F$ denote islands, and the lines connecting them bridges. A man begins at A and walks from island to island. He stops for lunch when he cannot continue to walk without crossing the same bridge twice. Find the number of ways that he can take his walk before eating lunch.

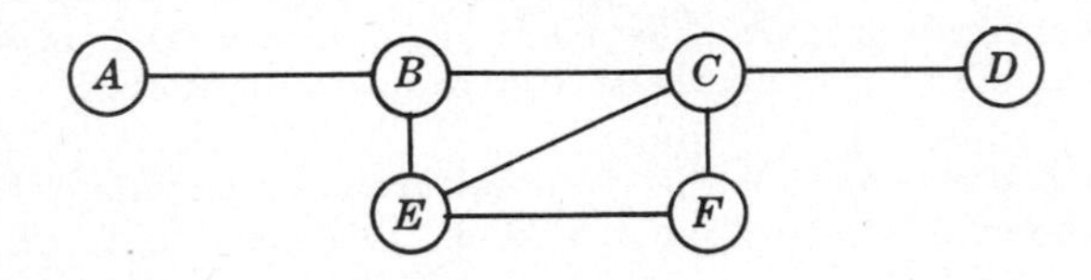

2.74. Consider the adjacent diagram with nine points $A, B, C, R, S, T, X, Y, Z$. A man begins at X and is allowed to move horizontally or vertically, one step at a time. He stops when he cannot continue to walk without reaching the same point more than once. Find the number of ways he can take his walk, if he first moves from X to R. (By symmetry, the total number of ways is twice this.)

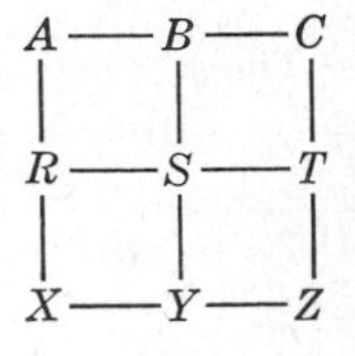

Answers to Supplementary Problems

2.31. (i) 362,880 (ii) 3,628,800 (iii) 39,916,800

2.32. (i) 240 (ii) 2184 (iii) 1/90 (iv) 1/1716

2.33. (i) $n+1$ (ii) $n(n-1) = n^2 - n$ (iii) $1/[n(n+1)(n+2)]$ (iv) $(n-r)(n-r+1)$

2.34. (i) $26 \cdot 25 \cdot 10 \cdot 9 \cdot 8 = 468{,}000$ (ii) $26 \cdot 25 \cdot 9 \cdot 9 \cdot 8 = 421{,}200$

2.35. (i) $6 \cdot 4 = 24$ (ii) $6 \cdot 4 \cdot 4 \cdot 6 = 24 \cdot 24 = 576$ (iii) $6 \cdot 4 \cdot 3 \cdot 5 = 360$

2.36. $3 \cdot 5 \cdot 4 \cdot 3 \cdot 2 \cdot 1 = 360$

2.37. (i) $5! = 120$ (ii) $4 \cdot 2! \cdot 3! = 48$

2.38. (i) $4! = 24$ (ii) $2!\,3! = 12$

2.39. (i) $7 \cdot 6 \cdot 5 \cdot 4 = 840$ (ii) $5 \cdot 4 \cdot 3 \cdot 2 = 120$ (iii) $5 \cdot 5 \cdot 4 \cdot 4 = 400$ (iv) $2 \cdot 6 \cdot 5 \cdot 4 = 240$ (v) $4 \cdot 6 \cdot 5 \cdot 4 = 480$ (vi) $1 \cdot 5 \cdot 4 \cdot 2 = 40$ (vii) $1 \cdot 3 \cdot 5 \cdot 4 = 60$ (viii) $4 \cdot 3 \cdot 5 \cdot 4 = 240$

2.40. $\dfrac{8!}{4!\,2!\,2!} = 420$

2.41. (i) $\dfrac{5!}{2!\,2!} = 30$ (ii) $\dfrac{9!}{2!\,2!\,2!} = 45{,}360$ (iii) $\dfrac{11!}{2!\,3!\,2!} = 1{,}663{,}200$ (iv) $\dfrac{8!}{2!\,2!\,2!} = 5040$

2.42. (i) $2 \cdot 4! \cdot 4! = 1152$ (ii) $2 \cdot 7 \cdot 3! \cdot 3! = 504$ (iii) $1152 - 504 = 648$

2.43. (i) $3! \cdot 4! = 144$ (ii) $2 \cdot 3! \cdot 3! = 72$ (iii) $144 - 72 = 72$

2.44. (i) $10 \cdot 10 \cdot 10 = 1000$ (iii) $10 \cdot 10 \cdot 10 \cdot 10 = 10{,}000$
(ii) $10 \cdot 9 \cdot 8 = 720$ (iv) $10 \cdot 9 \cdot 8 \cdot 7 \cdot 6 = 30{,}240$

2.45. $3!\,5!\,4!\,3! = 103{,}680$

2.46. (i) $3 \cdot 9 \cdot 8 = 216$ (ii) $8 \cdot 8 \cdot 5 = 320$
(iii) $9 \cdot 8 \cdot 1 = 72$ end in 0, and $8 \cdot 8 \cdot 4 = 256$ end in the other even digits; hence, altogether, $72 + 256 = 328$ are even.
(iv) $9 \cdot 8 \cdot 1 = 72$ end in 0, and $8 \cdot 8 \cdot 1 = 64$ end in 5; hence, altogether, $72 + 64 = 136$ are divisible by 5.

2.47. (i) $\frac{6!}{3!} = 120$ (ii) $4! = 24$ (iii) $4 \cdot 3! = 24$ (iv) $\frac{4!}{2!} = 12$

2.48. (i) 10 (ii) 35 (iii) 91 (iv) 15 (v) 1140 (vi) 816

2.49. (i) 504 (ii) 210 (iii) 180

2.50. (i) $8x^3 + 12x^2y^2 + 6xy^4 + y^6$
(ii) $x^8 - 12x^6y + 54x^4y^2 - 108x^2y^3 + 81y^4$
(iii) $a^5/32 + 5a^4b/8 + 5a^3b^2 + 20a^2b^3 + 40ab^4 + 32b^5$
(iv) $64a^{12} - 192a^{10}b + 240a^8b^2 - 160a^6b^3 + 60a^4b^4 - 12a^2b^5 + b^6$

2.51. *Hint.* Expand $(1+1)^n$. **2.53.** $70x^8y^{12}$

2.52. *Hint.* Expand $(1-1)^n$. **2.54.** $945x^3y^6z^8$

2.55. (i) $\binom{12}{4} = 495$, (ii) $\binom{12}{4} - \binom{9}{4} = 369$, (iii) $3 \cdot \binom{9}{3} = 252$

2.56. (i) $\binom{11}{5} = 462$, (ii) $\binom{9}{3} + \binom{9}{5} = 210$, (iii) $\binom{9}{5} + 2 \cdot \binom{9}{4} = 378$

2.57. (i) $\binom{10}{2} = 45$, (ii) $\binom{8}{2} = 28$, (iii) $\binom{10}{3} = 120$, (iv) $\binom{9}{2} = 36$, (v) 8

2.58. (i) $\binom{13}{10} = \binom{13}{3} = 286$ (iv) $\binom{5}{3}\binom{8}{7} = 80$
(ii) $\binom{11}{8} = \binom{11}{3} = 165$ (v) $\binom{5}{3}\binom{8}{7} + \binom{5}{4}\binom{8}{6} + \binom{5}{5}\binom{8}{5} = 276$
(iii) $2 \cdot \binom{11}{9} = 2 \cdot \binom{11}{2} = 110$

2.59. (i) $4 \cdot 10 = 40$, (ii) $13 \cdot 48 = 624$, (iii) $10 \cdot 4^5 - 40 = 10{,}200$. (We subtract the number of straight flushes.) (iv) $\binom{4}{2}\binom{12}{3} \cdot 4^3 = 84{,}480$, (v) $13 \cdot \binom{4}{2}\binom{12}{3} \cdot 4^3 = 1{,}098{,}240$

2.60. (i) $\binom{21}{3}\binom{5}{2} \cdot 5! = 1{,}596{,}000$ (v) $19 \cdot \binom{5}{2} \cdot 3! = 1140$ (ix) $4 \cdot \binom{20}{2} \cdot 3! = 4560$
(ii) $\binom{20}{2}\binom{5}{2} \cdot 5! = 228{,}000$ (vi) $4 \cdot \binom{20}{2} \cdot 5! = 91{,}200$ (x) $4 \cdot 19 \cdot 5! = 9120$
(iii) $19 \cdot \binom{5}{2} \cdot 5! = 22{,}800$ (vii) $4 \cdot \binom{20}{2} \cdot 4! = 18{,}240$
(iv) $19 \cdot \binom{5}{2} \cdot 4! = 4560$ (viii) 18,240 (same as (vii))

2.61. $\dfrac{9!}{3!\,3!\,3!} = 1680$

2.62. $1680/3! = 280$ or $\dbinom{8}{2}\dbinom{5}{2} = 280$

2.63. $\dfrac{10!}{4!\,3!\,3!} \cdot \dfrac{1}{2!} = 2100$ or $\dbinom{10}{4}\dbinom{5}{2} = 2100$

2.64. $\dfrac{12!}{3!\,3!\,3!\,3!} = 369{,}600$

2.65. $\dfrac{12!}{5!\,4!\,3!} = 27{,}720$

2.66. $2^{n-1} - 1$

2.67. $\dfrac{14!}{3!\,3!\,2!\,2!\,2!\,2!} \cdot \dfrac{1}{2!\,4!} = 3{,}153{,}150$

2.69.

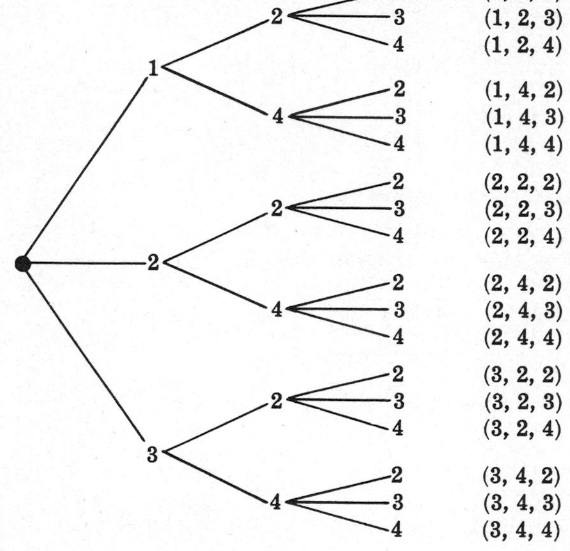

The eighteen elements of the product set are listed to the right of the tree diagram.

2.70. 14 ways

2.71. 20 ways (as seen in the following diagram):

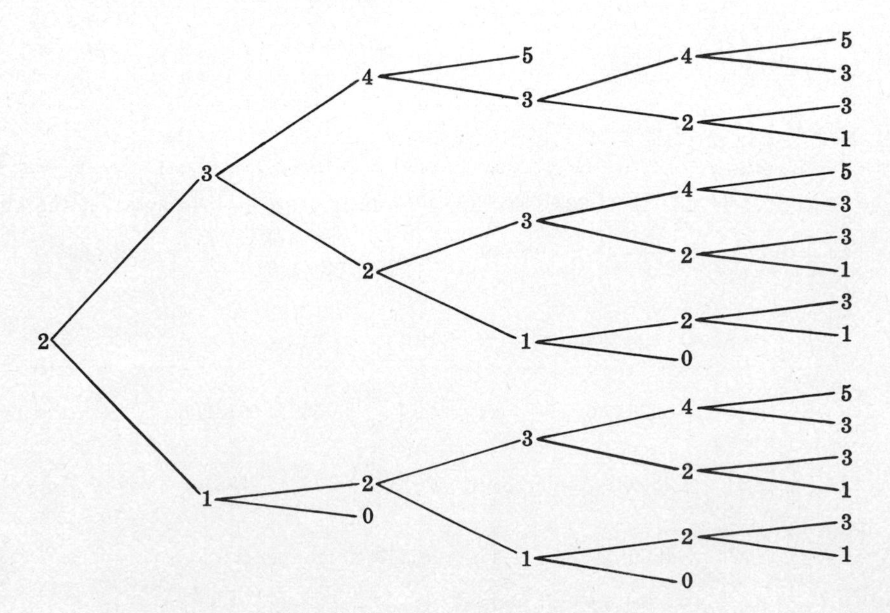

2.72. *Hint.* The tree is essentially the same as the tree of the preceding problem.

2.73. The appropriate tree diagram follows:

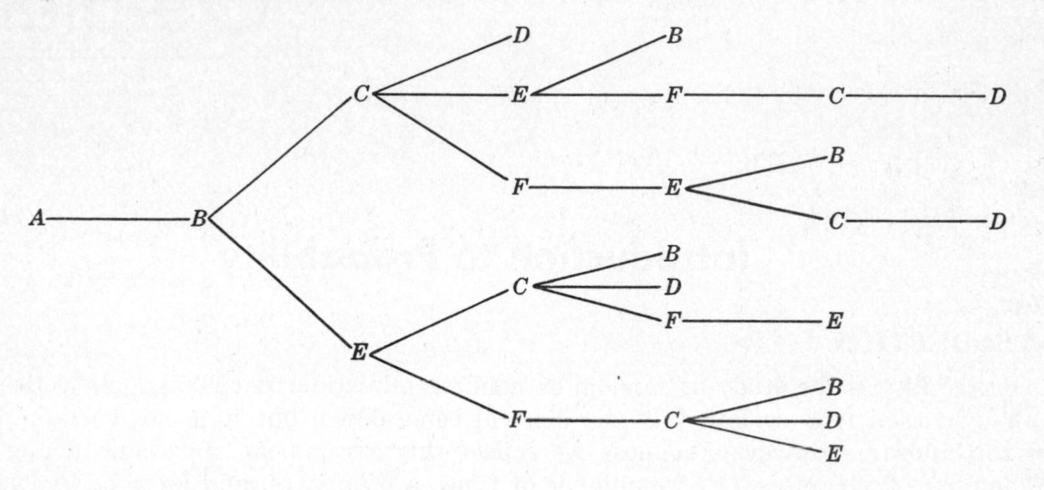

There are eleven ways to take his walk. Observe that he must eat his lunch at either B, D or E.

2.74. The appropriate tree diagram follows:

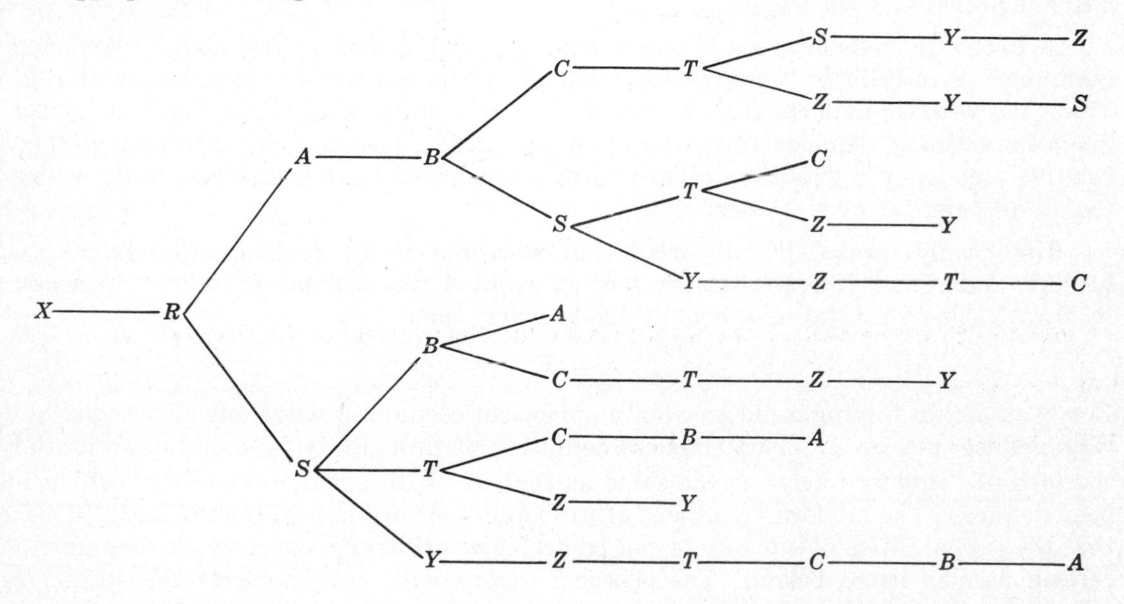

There are 10 different trips. (Note that in only 4 of them are all nine points covered.)

Chapter 3

Introduction to Probability

INTRODUCTION

Probability is the study of random or nondeterministic experiments. If a die is tossed in the air, then it is certain that the die will come down, but it is not certain that, say, a 6 will appear. However, suppose we repeat this experiment of tossing a die; let s be the number of successes, i.e. the number of times a 6 appears, and let n be the number of tosses. Then it has been empirically observed that the ratio $f = s/n$, called the *relative frequency,* becomes stable in the long run, i.e. approaches a limit. This stability is the basis of probability theory.

In probability theory, we define a mathematical model of the above phenomenon by assigning "probabilities" (or: the limit values of the relative frequencies) to the "events" connected with an experiment. Naturally, the reliability of our mathematical model for a given experiment depends upon the closeness of the assigned probabilities to the actual relative frequency. This then gives rise to problems of testing and reliability which form the subject matter of statistics.

Historically, probability theory began with the study of games of chance, such as roulette and cards. The probability p of an event A was defined as follows: if A can occur in s ways out of a total of n equally likely ways, then

$$p = P(A) = \frac{s}{n}$$

For example, in tossing a die an even number can occur in 3 ways out of 6 "equally likely" ways; hence $p = \frac{3}{6} = \frac{1}{2}$. This classical definition of probability is essentially circular since the idea of "equally likely" is the same as that of "with equal probability" which has not been defined. The modern treatment of probability theory is purely axiomatic. This means that the probabilities of our events can be perfectly arbitrary, except that they must satisfy certain axioms listed below. The classical theory will correspond to the special case of so-called *equiprobable spaces.*

SAMPLE SPACE AND EVENTS

The set S of all possible outcomes of some given experiment is called the *sample space.* A particular outcome, i.e. an element in S, is called a *sample point* or *sample.* An *event* A is a set of outcomes or, in other words, a subset of the sample space S. The event $\{a\}$ consisting of a single sample $a \in S$ is called an *elementary event.* The empty set $\emptyset$ and S itself are events; $\emptyset$ is sometimes called the *impossible* event, and S the *certain* or *sure* event.

We can combine events to form new events using the various set operations:

(i) $A \cup B$ is the event that occurs iff A occurs *or* B occurs (or both);

(ii) $A \cap B$ is the event that occurs iff A occurs *and* B occurs;

(iii) A^c, the complement of A, is the event that occurs iff A does *not* occur.

Two events A and B are called *mutually exclusive* if they are disjoint, i.e. if $A \cap B = \emptyset$. In other words, A and B are mutually exclusive if they cannot occur simultaneously.

Example 3.1: Experiment: Toss a die and observe the number that appears on top. Then the sample space consists of the six possible numbers:

$$S = \{1, 2, 3, 4, 5, 6\}$$

Let A be the event that an even number occurs, B that an odd number occurs and C that a prime number occurs:

$$A = \{2, 4, 6\}, \quad B = \{1, 3, 5\}, \quad C = \{2, 3, 5\}$$

Then:

$A \cup C = \{2, 3, 4, 5, 6\}$ is the event that an even or a prime number occurs;

$B \cap C = \{3, 5\}$ is the event that an odd prime number occurs;

$C^c = \{1, 4, 6\}$ is the event that a prime number does not occur.

Note that A and B are mutually exclusive: $A \cap B = \emptyset$; in other words, an even number and an odd number cannot occur simultaneously.

Example 3.2: Experiment: Toss a coin 3 times and observe the sequence of heads (H) and tails (T) that appears. The sample space S consists of eight elements:

$$S = \{\text{HHH, HHT, HTH, HTT, THH, THT, TTH, TTT}\}$$

Let A be the event that two or more heads appear consecutively, and B that all the tosses are the same:

$$A = \{\text{HHH, HHT, THH}\} \quad \text{and} \quad B = \{\text{HHH, TTT}\}$$

Then $A \cap B = \{\text{HHH}\}$ is the elementary event in which only heads appear. The event that 5 heads appear is the empty set $\emptyset$.

Example 3.3: Experiment: Toss a coin until a head appears and then count the number of times the coin was tossed. The sample space of this experiment is $S = \{1, 2, 3, \ldots, \infty\}$. Here ∞ refers to the case when a head never appears and so the coin is tossed an infinite number of times. This is an example of a sample space which is *countably infinite*.

Example 3.4: Experiment: Let a pencil drop, head first, into a rectangular box and note the point on the bottom of the box that the pencil first touches. Here S consists of all the points on the bottom of the box. Let the rectangular area on the right represent these points. Let A and B be the events that the pencil drops into the corresponding areas illustrated on the right. This is an example of a sample space which is not finite nor even countably infinite, i.e. which is uncountable.

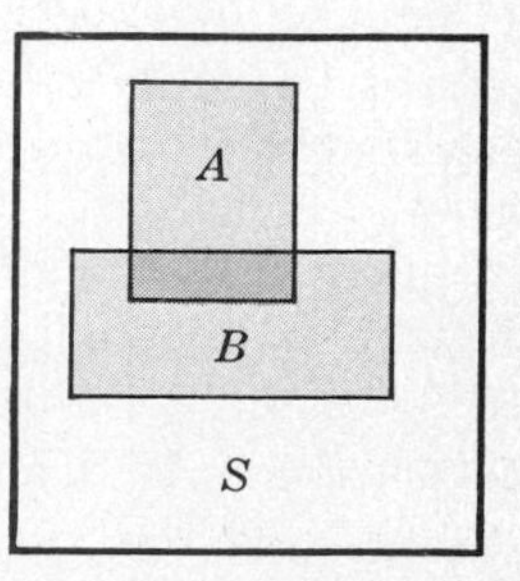

Remark: If the sample space S is finite or countably infinite, then every subset of S is an event. On the other hand, if S is uncountable, as in Example 3.4, then for technical reasons (which lie beyond the scope of this text) some subsets of S cannot be events. However, in all cases the events shall form a σ-algebra $\mathcal{E}$ of subsets of S.

AXIOMS OF PROBABILITY

Let S be a sample space, let $\mathcal{E}$ be the class of events, and let P be a real-valued function defined on $\mathcal{E}$. Then P is called a *probability function*, and $P(A)$ is called the *probability* of the event A if the following axioms hold:

[$\mathbf{P}_1$] For every event A, $0 \leq P(A) \leq 1$.

[$\mathbf{P}_2$] $P(S) = 1$.

[$\mathbf{P}_3$] If A and B are mutually exclusive events, then

$$P(A \cup B) = P(A) + P(B)$$

[$\mathbf{P}_4$] If $A_1, A_2, \ldots$ is a sequence of mutually exclusive events, then

$$P(A_1 \cup A_2 \cup \cdots) = P(A_1) + P(A_2) + \cdots$$

The following remarks concerning the axioms [$\mathbf{P}_3$] and [$\mathbf{P}_4$] are in order. First of all, using [$\mathbf{P}_3$] and mathematical induction we can prove that for any mutually exclusive events $A_1, A_2, \ldots, A_n$,

$$P(A_1 \cup A_2 \cup \cdots \cup A_n) = P(A_1) + P(A_2) + \cdots + P(A_n) \qquad (*)$$

We emphasize that [$\mathbf{P}_4$] does not follow from [$\mathbf{P}_3$] even though (*) holds for every positive integer n. However, if the sample space S is finite, then clearly the axiom [$\mathbf{P}_4$] is superfluous.

We now prove a number of theorems which follow directly from our axioms.

Theorem 3.1: If $\emptyset$ is the empty set, then $P(\emptyset) = 0$.

Proof: Let A be any set; then A and $\emptyset$ are disjoint and $A \cup \emptyset = A$. By [$\mathbf{P}_3$],

$$P(A) = P(A \cup \emptyset) = P(A) + P(\emptyset)$$

Subtracting $P(A)$ from both sides gives our result.

Theorem 3.2: If A^c is the complement of an event A, then $P(A^c) = 1 - P(A)$.

Proof: The sample space S can be decomposed into the mutually exclusive events A and A^c; that is, $S = A \cup A^c$. By [$\mathbf{P}_2$] and [$\mathbf{P}_3$] we obtain

$$1 = P(S) = P(A \cup A^c) = P(A) + P(A^c)$$

from which our result follows.

Theorem 3.3: If $A \subset B$, then $P(A) \leq P(B)$.

Proof. If $A \subset B$, then B can be decomposed into the mutually exclusive events A and $B \setminus A$ (as illustrated on the right). Thus

$$P(B) = P(A) + P(B \setminus A)$$

The result now follows from the fact that $P(B \setminus A) \geq 0$.

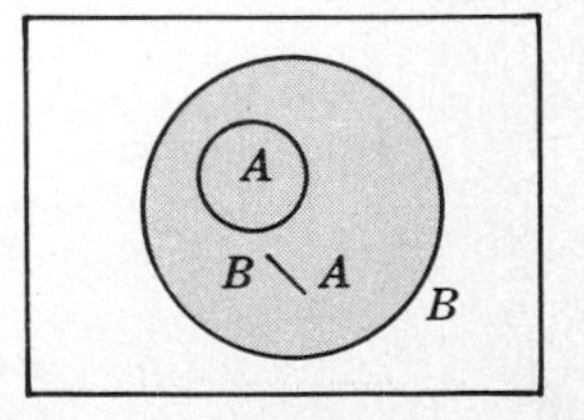

B is shaded.

Theorem 3.4: If A and B are any two events, then

$$P(A \setminus B) = P(A) - P(A \cap B)$$

Proof. Now A can be decomposed into the mutually exclusive events $A \setminus B$ and $A \cap B$; that is, $A = (A \setminus B) \cup (A \cap B)$. Thus by [$\mathbf{P}_3$],

$$P(A) = P(A \setminus B) + P(A \cap B)$$

from which our result follows.

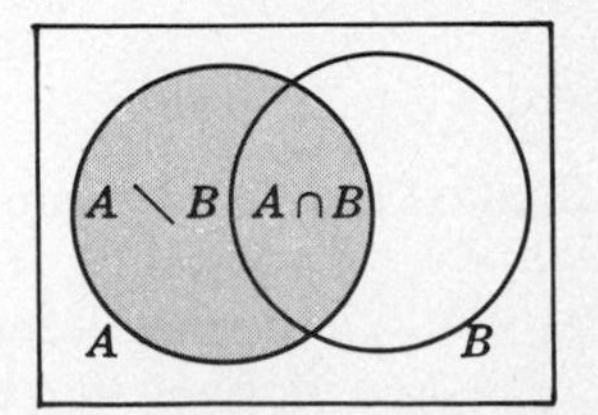

A is shaded.

Theorem 3.5: If A and B are any two events, then

$$P(A \cup B) = P(A) + P(B) - P(A \cap B)$$

Proof. Note that $A \cup B$ can be decomposed into the mutually exclusive events $A \setminus B$ and B; that is, $A \cup B = (A \setminus B) \cup B$. Thus by [$\mathbf{P}_3$] and Theorem 3.4,

$$\begin{aligned} P(A \cup B) &= P(A \setminus B) + P(B) \\ &= P(A) - P(A \cap B) + P(B) \\ &= P(A) + P(B) - P(A \cap B) \end{aligned}$$

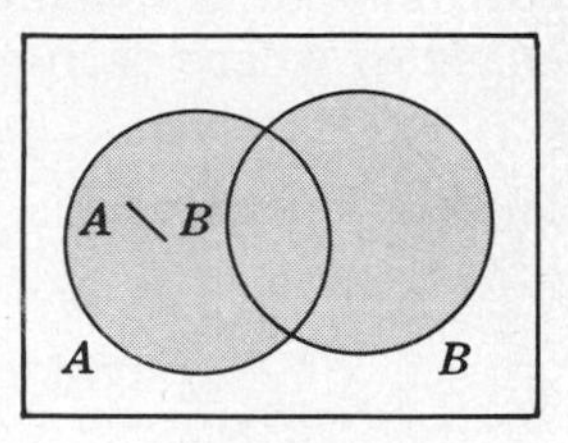

$A \cup B$ is shaded.

which is the desired result.

Applying the above theorem twice (Problem 3.23) we obtain

Corollary 3.6: For any events A, B and C,

$$P(A \cup B \cup C) = P(A) + P(B) + P(C) - P(A \cap B) - P(A \cap C) - P(B \cap C) + P(A \cap B \cap C)$$

FINITE PROBABILITY SPACES

Let S be a finite sample space; say, $S = \{a_1, a_2, \ldots, a_n\}$. A finite probability space is obtained by assigning to each point $a_i \in S$ a real number p_i, called the *probability* of a_i, satisfying the following properties:

(i) each p_i is nonnegative, $p_i \geq 0$

(ii) the sum of the p_i is one, $p_1 + p_2 + \cdots + p_n = 1$.

The *probability* $P(A)$ of any event A, is then defined to be the sum of the probabilities of the points in A. For notational convenience we write $P(a_i)$ for $P(\{a_i\})$.

Example 3.5: Let three coins be tossed and the number of heads observed; then the sample space is $S = \{0, 1, 2, 3\}$. We obtain a probability space by the following assignment

$$P(0) = \tfrac{1}{8}, \quad P(1) = \tfrac{3}{8}, \quad P(2) = \tfrac{3}{8} \quad \text{and} \quad P(3) = \tfrac{1}{8}$$

since each probability is nonnegative and the sum of the probabilities is 1. Let A be the event that at least one head appears and let B be the event that all heads or all tails appear:

$$A = \{1, 2, 3\} \quad \text{and} \quad B = \{0, 3\}$$

Then, by definition,

$$P(A) = P(1) + P(2) + P(3) = \tfrac{3}{8} + \tfrac{3}{8} + \tfrac{1}{8} = \tfrac{7}{8}$$

and
$$P(B) = P(0) + P(3) = \tfrac{1}{8} + \tfrac{1}{8} = \tfrac{1}{4}$$

Example 3.6: Three horses A, B and C are in a race; A is twice as likely to win as B and B is twice as likely to win as C. What are their respective probabilities of winning, i.e. $P(A)$, $P(B)$ and $P(C)$?

Let $P(C) = p$; since B is twice as likely to win as C, $P(B) = 2p$; and since A is twice as likely to win as B, $P(A) = 2P(B) = 2(2p) = 4p$. Now the sum of the probabilities must be 1; hence

$$p + 2p + 4p = 1 \quad \text{or} \quad 7p = 1 \quad \text{or} \quad p = \tfrac{1}{7}$$

Accordingly,

$$P(A) = 4p = \tfrac{4}{7}, \quad P(B) = 2p = \tfrac{2}{7}, \quad P(C) = p = \tfrac{1}{7}$$

Question: What is the probability that B or C wins, i.e. $P(\{B, C\})$? By definition

$$P(\{B, C\}) = P(B) + P(C) = \tfrac{2}{7} + \tfrac{1}{7} = \tfrac{3}{7}$$

FINITE EQUIPROBABLE SPACES

Frequently, the physical characteristics of an experiment suggest that the various outcomes of the sample space be assigned equal probabilities. Such a finite probability space S, where each sample point has the same probability, will be called an *equiprobable* or *uniform space*. In particular, if S contains n points then the probability of each point is $1/n$. Furthermore, if an event A contains r points then its probability is $r \cdot \frac{1}{n} = \frac{r}{n}$. In other words,

$$P(A) = \frac{\text{number of elements in } A}{\text{number of elements in } S}$$

or

$$P(A) = \frac{\text{number of ways that the event } A \text{ can occur}}{\text{number of ways that the sample space } S \text{ can occur}}$$

We emphasize that the above formula for $P(A)$ can only be used with respect to an equiprobable space, and cannot be used in general.

The expression "at random" will be used only with respect to an equiprobable space; formally, the statement "choose a point at random from a set S" shall mean that S is an equiprobable space, i.e. that each sample point in S has the same probability.

Example 3.7: Let a card be selected at random from an ordinary deck of 52 cards. Let

$$A = \{\text{the card is a spade}\}$$

and

$$B = \{\text{the card is a face card, i.e. a jack, queen or king}\}$$

We compute $P(A)$, $P(B)$ and $P(A \cap B)$. Since we have an equiprobable space,

$$P(A) = \frac{\text{number of spades}}{\text{number of cards}} = \frac{13}{52} = \frac{1}{4} \qquad P(B) = \frac{\text{number of face cards}}{\text{number of cards}} = \frac{12}{52} = \frac{3}{13}$$

$$P(A \cap B) = \frac{\text{number of spade face cards}}{\text{number of cards}} = \frac{3}{52}$$

Example 3.8: Let 2 items be chosen at random from a lot containing 12 items of which 4 are defective. Let

$$A = \{\text{both items are defective}\} \quad \text{and} \quad B = \{\text{both items are non-defective}\}$$

Find $P(A)$ and $P(B)$. Now

S can occur in $\binom{12}{2}$ = 66 ways, the number of ways that 2 items can be chosen from 12 items;

A can occur in $\binom{4}{2}$ = 6 ways, the number of ways that 2 defective items can be chosen from 4 defective items;

B can occur in $\binom{8}{2}$ = 28 ways, the number of ways that 2 non-defective items can be chosen from 8 non-defective items.

Accordingly, $P(A) = \frac{6}{66} = \frac{1}{11}$ and $P(B) = \frac{28}{66} = \frac{14}{33}$.

Question: What is the probability that at least one item is defective? Now

$$C = \{\text{at least one item is defective}\}$$

is the complement of B; that is, $C = B^c$. Thus by Theorem 3.2,

$$P(C) = P(B^c) = 1 - P(B) = 1 - \tfrac{14}{33} = \tfrac{19}{33}$$

The *odds* that an event with probability p occurs is defined to be the ratio $p : (1-p)$. Thus the odds that at least one item is defective is $\frac{19}{33} : \frac{14}{33}$ or $19:14$ which is read "19 to 14".

Example 3.9: (Classical Birthday Problem.) We seek the probability p that n people have distinct birthdays. In solving this problem, we ignore leap years and assume that a person's birthday can fall on any day with the same probability.

Since there are n people and 365 different days, there are 365^n ways in which the n people can have their birthdays. On the other hand, if the n persons are to have distinct birthdays, then the first person can be born on any of the 365 days, the second person can be born on the remaining 364 days, the third person can be born on the remaining 363 days, etc. Thus there are $365 \cdot 364 \cdot 363 \cdots (365-n+1)$ ways the n persons can have distinct birthdays. Accordingly,

$$p = \frac{365 \cdot 364 \cdot 363 \cdots (365-n+1)}{365^n} = \frac{365}{365} \cdot \frac{364}{365} \cdot \frac{363}{365} \cdots \frac{365-n+1}{365}$$

It can be shown that for $n \geqq 23$, $p < \frac{1}{2}$; in other words, amongst 23 or more people it is more likely that at least two of them have the same birthday than that they all have distinct birthdays.

INFINITE SAMPLE SPACES

Now suppose S is a countably infinite sample space; say $S = \{a_1, a_2, \ldots\}$. As in the finite case, we obtain a probability space by assigning to each $a_i \in S$ a real number p_i, called its probability, such that

$$\text{(i)}\quad p_i \geqq 0 \quad \text{and} \quad \text{(ii)}\quad p_1 + p_2 + \cdots = \sum_{i=1}^{\infty} p_i = 1$$

The probability $P(A)$ of any event A is then the sum of the probabilities of its points.

Example 3.10: Consider the sample space $S = \{1, 2, 3, \ldots, \infty\}$ of the experiment of tossing a coin till a head appears; here n denotes the number of times the coin is tossed. A probability space is obtained by setting

$$p(1) = \tfrac{1}{2}, \quad p(2) = \tfrac{1}{4}, \quad \ldots, \quad p(n) = 1/2^n, \quad \ldots, \quad p(\infty) = 0$$

The only uncountable sample spaces S which we will consider here are those with some finite geometrical measurement $m(S)$ such as length, area or volume, and in which a point is selected at random. The probability of an event A, i.e. that the selected point belongs to A, is then the ratio of $m(A)$ to $m(S)$; that is,

$$P(A) = \frac{\text{length of } A}{\text{length of } S} \quad \text{or} \quad P(A) = \frac{\text{area of } A}{\text{area of } S} \quad \text{or} \quad P(A) = \frac{\text{volume of } A}{\text{volume of } S}$$

Such a probability space is said to be *uniform*.

Example 3.11: On the real line $\mathbf{R}$, points a and b are selected at random such that $-2 \leqq b \leqq 0$ and $0 \leqq a \leqq 3$, as shown below. Find the probability p that the distance d between a and b is greater than 3.

The sample space S consists of the ordered pairs (a, b) and so forms the rectangular region shown in the adjacent diagram. On the other hand, the set A of points (a, b) for which $d = a - b > 3$ consists of those points of S which lie below the line $x - y = 3$, and hence forms the shaded area in the diagram. Thus

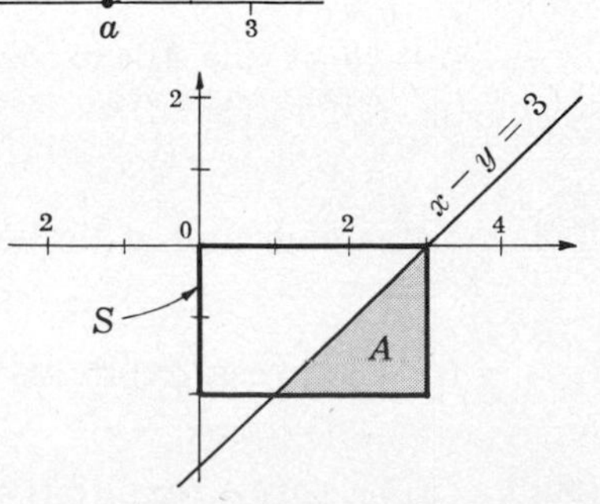

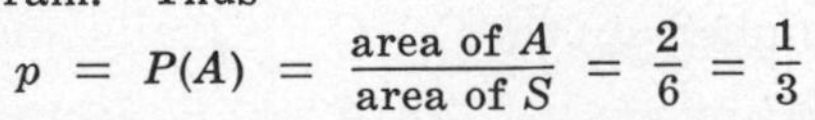

$$p = P(A) = \frac{\text{area of } A}{\text{area of } S} = \frac{2}{6} = \frac{1}{3}$$

Remark: A finite or countably infinite probability space is said to be *discrete*, and an uncountable space is said to be *nondiscrete*.

Solved Problems

SAMPLE SPACES AND EVENTS

3.1. Let A and B be events. Find an expression and exhibit the Venn diagram for the event that: (i) A but not B occurs, i.e. only A occurs; (ii) either A or B, but not both, occurs, i.e. exactly one of the two events occurs.

(i) Since A but not B occurs, shade the area of A outside of B as in Figure (a) below. Note that B^c, the complement of B, occurs since B does not occur; hence A and B^c occurs. In other words, the event is $A \cap B^c$.

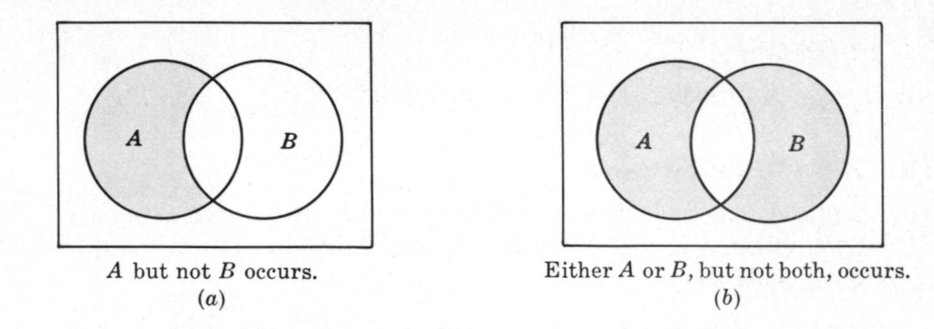

A but not B occurs.
(a)

Either A or B, but not both, occurs.
(b)

(ii) Since A or B but not both occurs, shade the area of A and B except where they intersect as in Figure (b) above. The event is equivalent to A but not B occurs or B but not A occurs. Now, as in (i), A but not B is the event $A \cap B^c$, and B but not A is the event $B \cap A^c$. Thus the given event is $(A \cap B^c) \cup (B \cap A^c)$.

3.2. Let A, B and C be events. Find an expression and exhibit the Venn diagram for the event that (i) A and B but not C occurs, (ii) only A occurs.

(i) Since A and B but not C occurs, shade the intersection of A and B which lies outside of C, as in Figure (a) below. The event is $A \cap B \cap C^c$.

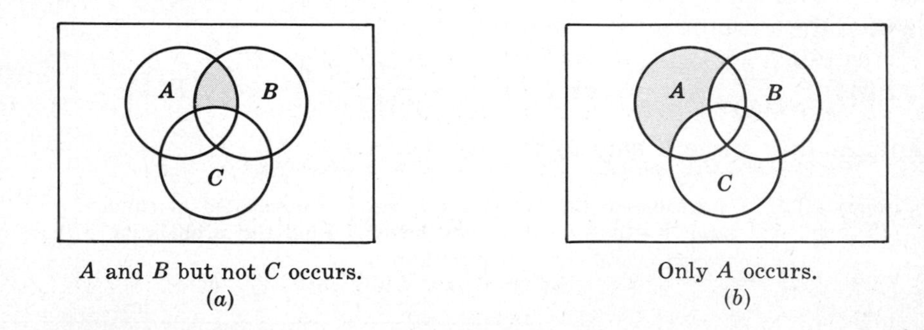

A and B but not C occurs.
(a)

Only A occurs.
(b)

(ii) Since only A is to occur, shade the area of A which lies outside of B and of C, as in Figure (b) above. The event is $A \cap B^c \cap C^c$.

3.3. Let a coin and a die be tossed; let the sample space S consist of the twelve elements:

$$S = \{\text{H1, H2, H3, H4, H5, H6, T1, T2, T3, T4, T5, T6}\}$$

(i) Express explicitly the following events: A = {heads and an even number appear}, B = {a prime number appears}, C = {tails and an odd number appear}.

(ii) Express explicitly the event that: (a) A or B occurs, (b) B and C occurs, (c) only B occurs.

(iii) Which of the events A, B and C are mutually exclusive?

(i) To obtain A, choose those elements of S consisting of an H and an even number: $A = \{H2, H4, H6\}$.

To obtain B, choose those points in S consisting of a prime number: $B = \{H2, H3, H5, T2, T3, T5\}$.

To obtain C, choose those points in S consisting of a T and an odd number: $C = \{T1, T3, T5\}$.

(ii) (a) A or $B = A \cup B = \{H2, H4, H6, H3, H5, T2, T3, T5\}$

(b) B and $C = B \cap C = \{T3, T5\}$

(c) Choose those elements of B which do not lie in A or C: $B \cap A^c \cap C^c = \{H3, H5, T2\}$.

(iii) A and C are mutually exclusive since $A \cap C = \emptyset$.

FINITE PROBABILITY SPACES

3.4. Suppose a sample space S consists of 4 elements: $S = \{a_1, a_2, a_3, a_4\}$. Which function defines a probability space on S?

(i) $P(a_1) = \frac{1}{2},\ P(a_2) = \frac{1}{3},\ P(a_3) = \frac{1}{4},\ P(a_4) = \frac{1}{5}$.

(ii) $P(a_1) = \frac{1}{2},\ P(a_2) = \frac{1}{4},\ P(a_3) = -\frac{1}{4},\ P(a_4) = \frac{1}{2}$.

(iii) $P(a_1) = \frac{1}{2},\ P(a_2) = \frac{1}{4},\ P(a_3) = \frac{1}{8},\ P(a_4) = \frac{1}{8}$.

(iv) $P(a_1) = \frac{1}{2},\ P(a_2) = \frac{1}{4},\ P(a_3) = \frac{1}{4},\ P(a_4) = 0$.

(i) Since the sum of the values on the sample points is greater than one, $\frac{1}{2} + \frac{1}{3} + \frac{1}{4} + \frac{1}{5} = \frac{77}{60}$, the function does not define a probability space on S.

(ii) Since $P(a_3) = -\frac{1}{4}$, a negative number, the function does not define a probability space on S.

(iii) Since each value is nonnegative, and the sum of the values is one, $\frac{1}{2} + \frac{1}{4} + \frac{1}{8} + \frac{1}{8} = 1$, the function does define a probability space on S.

(iv) The values are nonnegative and add up to one; hence the function does define a probability space on S.

3.5. Let $S = \{a_1, a_2, a_3, a_4\}$, and let P be a probability function on S.

(i) Find $P(a_1)$ if $P(a_2) = \frac{1}{3},\ P(a_3) = \frac{1}{6},\ P(a_4) = \frac{1}{9}$.

(ii) Find $P(a_1)$ and $P(a_2)$ if $P(a_3) = P(a_4) = \frac{1}{4}$ and $P(a_1) = 2P(a_2)$.

(iii) Find $P(a_1)$ if $P(\{a_2, a_3\}) = \frac{2}{3},\ P(\{a_2, a_4\}) = \frac{1}{2}$ and $P(a_2) = \frac{1}{3}$.

(i) Let $P(a_1) = p$. Then for P to be a probability function, the sum of the probabilities on the sample points must be one: $p + \frac{1}{3} + \frac{1}{6} + \frac{1}{9} = 1$ or $p = \frac{7}{18}$.

(ii) Let $P(a_2) = p$, then $P(a_1) = 2p$. Hence $2p + p + \frac{1}{4} + \frac{1}{4} = 1$ or $p = \frac{1}{6}$. Thus $P(a_2) = \frac{1}{6}$ and $P(a_1) = \frac{1}{3}$.

(iii) Let $P(a_1) = p$.

$$P(a_3) = P(\{a_2, a_3\}) - P(a_2) = \tfrac{2}{3} - \tfrac{1}{3} = \tfrac{1}{3}$$
$$P(a_4) = P(\{a_2, a_4\}) - P(a_2) = \tfrac{1}{2} - \tfrac{1}{3} = \tfrac{1}{6}$$

Then $p + \frac{1}{3} + \frac{1}{3} + \frac{1}{6} = 1$ or $p = \frac{1}{6}$, that is, $P(a_1) = \frac{1}{6}$.

3.6. A coin is weighted so that heads is twice as likely to appear as tails. Find $P(T)$ and $P(H)$.

Let $P(T) = p$; then $P(H) = 2p$. Now set the sum of the probabilities equal to one: $p + 2p = 1$ or $p = \frac{1}{3}$. Thus $P(T) = p = \frac{1}{3}$ and $P(H) = 2p = \frac{2}{3}$.

3.7. Two men, m_1 and m_2, and three women, w_1, w_2 and w_3, are in a chess tournament. Those of the same sex have equal probabilities of winning, but each man is twice as likely to win as any woman. (i) Find the probability that a woman wins the tournament. (ii) If m_1 and w_1 are married, find the probability that one of them wins the tournament.

Set $P(w_1) = p$; then $P(w_2) = P(w_3) = p$ and $P(m_1) = P(m_2) = 2p$. Next set the sum of the probabilities of the five sample points equal to one: $p + p + p + 2p + 2p = 1$ or $p = \frac{1}{7}$.

We seek (i) $P(\{w_1, w_2, w_3\})$ and (ii) $P(\{m_1, w_1\})$. Then by definition,

$$P(\{w_1, w_2, w_3\}) = P(w_1) + P(w_2) + P(w_3) = \tfrac{1}{7} + \tfrac{1}{7} + \tfrac{1}{7} = \tfrac{3}{7}$$

$$P(\{m_1, w_1\}) = P(m_1) + P(w_1) = \tfrac{2}{7} + \tfrac{1}{7} = \tfrac{3}{7}$$

3.8. Let a die be weighted so that the probability of a number appearing when the die is tossed is proportional to the given number (e.g. 6 has twice the probability of appearing as 3). Let $A = \{\text{even number}\}$, $B = \{\text{prime number}\}$, $C = \{\text{odd number}\}$.

(i) Describe the probability space, i.e. find the probability of each sample point.

(ii) Find $P(A)$, $P(B)$ and $P(C)$.

(iii) Find the probability that: (*a*) an even or prime number occurs; (*b*) an odd prime number occurs; (*c*) A but not B occurs.

(i) Let $P(1) = p$. Then $P(2) = 2p$, $P(3) = 3p$, $P(4) = 4p$, $P(5) = 5p$ and $P(6) = 6p$. Since the sum of the probabilities must be one, we obtain $p + 2p + 3p + 4p + 5p + 6p = 1$ or $p = 1/21$. Thus

$$P(1) = \tfrac{1}{21},\quad P(2) = \tfrac{2}{21},\quad P(3) = \tfrac{1}{7},\quad P(4) = \tfrac{4}{21},\quad P(5) = \tfrac{5}{21},\quad P(6) = \tfrac{2}{7}$$

(ii) $P(A) = P(\{2,4,6\}) = \frac{4}{7}$, $P(B) = P(\{2,3,5\}) = \frac{10}{21}$, $P(C) = P(\{1,3,5\}) = \frac{3}{7}$.

(iii) (*a*) The event that an even or prime number occurs is $A \cup B = \{2,4,6,3,5\}$, or that 1 does not occur. Thus $P(A \cup B) = 1 - P(1) = \frac{20}{21}$.

(*b*) The event that an odd prime number occurs is $B \cap C = \{3,5\}$. Thus $P(B \cap C) = P(\{3,5\}) = \frac{8}{21}$.

(*c*) The event that A but not B occurs is $A \cap B^c = \{4,6\}$. Hence $P(A \cap B^c) = P(\{4,6\}) = \frac{10}{21}$.

FINITE EQUIPROBABLE SPACES

3.9. Determine the probability p of each event:

(i) an even number appears in the toss of a fair die;

(ii) a king appears in drawing a single card from an ordinary deck of 52 cards;

(iii) at least one tail appears in the toss of three fair coins;

(iv) a white marble appears in drawing a single marble from an urn containing 4 white, 3 red and 5 blue marbles.

(i) The event can occur in three ways (a 2, 4 or 6) out of 6 equally likely cases; hence $p = \frac{3}{6} = \frac{1}{2}$.

(ii) There are 4 kings among the 52 cards; hence $p = \frac{4}{52} = \frac{1}{13}$.

(iii) If we consider the coins distinguished, then there are 8 equally likely cases: HHH, HHT, HTH, HTT, THH, THT, TTH, TTT. Only the first case is not favorable to the given event; hence $p = \frac{7}{8}$.

(iv) There are $4 + 3 + 5 = 12$ marbles, of which 4 are white; hence $p = \frac{4}{12} = \frac{1}{3}$.

3.10. Two cards are drawn at random from an ordinary deck of 52 cards. Find the probability p that (i) both are spades, (ii) one is a spade and one is a heart.

There are $\binom{52}{2} = 1326$ ways to draw 2 cards from 52 cards.

(i) There are $\binom{13}{2} = 78$ ways to draw 2 spades from 13 spades; hence

$$p = \frac{\text{number of ways 2 spades can be drawn}}{\text{number of ways 2 cards can be drawn}} = \frac{78}{1326} = \frac{1}{17}$$

(ii) Since there are 13 spades and 13 hearts, there are $13 \cdot 13 = 169$ ways to draw a spade and a heart; hence $p = \frac{169}{1326} = \frac{13}{102}$.

3.11. Three light bulbs are chosen at random from 15 bulbs of which 5 are defective. Find the probability p that (i) none is defective, (ii) exactly one is defective, (iii) at least one is defective.

There are $\binom{15}{3} = 455$ ways to choose 3 bulbs from the 15 bulbs.

(i) Since there are $15 - 5 = 10$ nondefective bulbs, there are $\binom{10}{3} = 120$ ways to choose 3 nondefective bulbs. Thus $p = \frac{120}{455} = \frac{24}{91}$.

(ii) There are 5 defective bulbs and $\binom{10}{2} = 45$ different pairs of nondefective bulbs; hence there are $5 \cdot 45 = 225$ ways to choose 3 bulbs of which one is defective. Thus $p = \frac{225}{455} = \frac{45}{91}$.

(iii) The event that at least one is defective is the complement of the event that none are defective which has, by (i), probability $\frac{24}{91}$. Hence $p = 1 - \frac{24}{91} = \frac{67}{91}$.

3.12. Two cards are selected at random from 10 cards numbered 1 to 10. Find the probability p that the sum is odd if (i) the two cards are drawn together, (ii) the two cards are drawn one after the other without replacement, (iii) the two cards are drawn one after the other with replacement.

(i) There are $\binom{10}{2} = 45$ ways to select 2 cards out of 10. The sum is odd if one number is odd and the other is even. There are 5 even numbers and 5 odd numbers; hence there are $5 \cdot 5 = 25$ ways of choosing an even and an odd number. Thus $p = \frac{25}{45} = \frac{5}{9}$.

(ii) There are $10 \cdot 9 = 90$ ways to draw two cards one after the other without replacement. There are $5 \cdot 5 = 25$ ways to draw an even number and then an odd number, and $5 \cdot 5 = 25$ ways to draw an odd number and then an even number; hence $p = \frac{25+25}{90} = \frac{50}{90} = \frac{5}{9}$.

(iii) There are $10 \cdot 10 = 100$ ways to draw two cards one after the other with replacement. As in (ii), there are $5 \cdot 5 = 25$ ways to draw an even number and then an odd number, and $5 \cdot 5 = 25$ ways to draw an odd number and then an even number; hence $p = \frac{25+25}{100} = \frac{50}{100} = \frac{1}{2}$.

3.13. Six married couples are standing in a room.

(i) If 2 people are chosen at random, find the probability p that (*a*) they are married, (*b*) one is male and one is female.

(ii) If 4 people are chosen at random, find the probability p that (*a*) 2 married couples are chosen, (*b*) no married couple is among the 4, (*c*) exactly one married couple is among the 4.

(iii) If the 12 people are divided into six pairs, find the probability p that (*a*) each pair is married, (*b*) each pair contains a male and a female.

(i) There are $\binom{12}{2} = 66$ ways to choose 2 people from the 12 people.

(a) There are 6 married couples; hence $p = \frac{6}{66} = \frac{1}{11}$.

(b) There are 6 ways to choose a male and 6 ways to choose a female; hence $p = \frac{6 \cdot 6}{66} = \frac{6}{11}$.

(ii) There are $\binom{12}{4} = 495$ ways to choose 4 people from the 12 people.

(a) There are $\binom{6}{2} = 15$ ways to choose 2 couples from the 6 couples; hence $p = \frac{15}{495} = \frac{1}{33}$.

(b) The 4 persons come from 4 different couples. There are $\binom{6}{4} = 15$ ways to choose 4 couples from the 6 couples, and there are 2 ways to choose one person from each couple. Hence $p = \frac{2 \cdot 2 \cdot 2 \cdot 2 \cdot 15}{495} = \frac{16}{33}$.

(c) This event is mutually disjoint from the preceding two events (which are also mutually disjoint) and at least one of these events must occur. Hence $p + \frac{1}{33} + \frac{16}{33} = 1$ or $p = \frac{16}{33}$.

(iii) There are $\frac{12!}{2!\,2!\,2!\,2!\,2!\,2!} = \frac{12!}{2^6}$ ways to partition the 12 people into 6 ordered cells with 2 people in each.

(a) The 6 couples can be placed into the 6 ordered cells in 6! ways. Hence $p = \frac{6!}{12!/2^6} = \frac{1}{10{,}395}$.

(b) The six men can be placed one each into the 6 cells in 6! ways, and the 6 women can be placed one each into the 6 cells in 6! ways. Hence $p = \frac{6!\,6!}{12!/2^6} = \frac{16}{231}$.

3.14. A class contains 10 men and 20 women of which half the men and half the women have brown eyes. Find the probability p that a person chosen at random is a man or has brown eyes.

Let $A = \{\text{person is a man}\}$ and $B = \{\text{person has brown eyes}\}$. We seek $P(A \cup B)$.

Then $P(A) = \frac{10}{30} = \frac{1}{3}$, $P(B) = \frac{15}{30} = \frac{1}{2}$, $P(A \cap B) = \frac{5}{30} = \frac{1}{6}$. Thus by Theorem 3.5,

$$p = P(A \cup B) = P(A) + P(B) - P(A \cap B) = \tfrac{1}{3} + \tfrac{1}{2} - \tfrac{1}{6} = \tfrac{2}{3}$$

UNCOUNTABLE UNIFORM SPACES

3.15. A point is selected at random inside a circle. Find the probability p that the point is closer to the center of the circle than to its circumference.

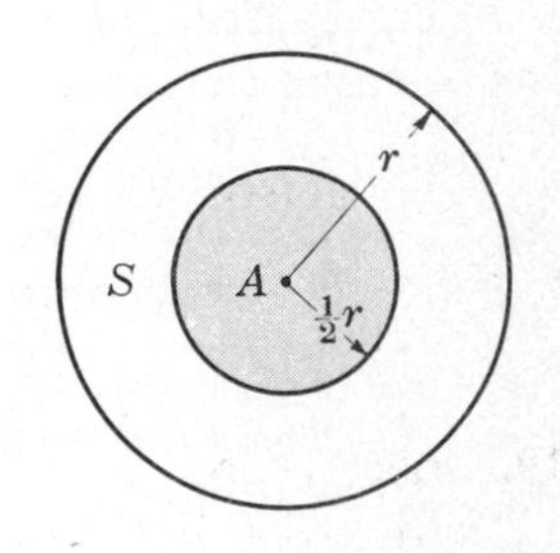

Let S denote the set of points inside the circle with radius r, and let A denote the set of points inside the concentric circle of radius $\frac{1}{2}r$. (Thus A consists precisely of those points of S which are closer to its center than to its circumference.) Accordingly,

$$p = P(A) = \frac{\text{area of } A}{\text{area of } S} = \frac{\pi(\frac{1}{2}r)^2}{\pi r^2} = \frac{1}{4}$$

3.16. Consider the Cartesian plane $\mathbf{R}^2$, and let X denote the subset of points for which both coordinates are integers. A coin of diameter $\frac{1}{2}$ is tossed randomly onto the plane. Find the probability p that the coin covers a point of X.

Let S denote the set of points inside a square with corners

$$(m, n), \quad (m, n+1), \quad (m+1, n), \quad (m+1, n+1) \in X$$

Let A denote the set of points in S with distance less than $\frac{1}{4}$ from any corner point. (Observe that the area of A is equal to the area inside a circle of radius $\frac{1}{4}$.) Thus a coin whose center falls in S will cover a point of X if and only if its center falls in a point of A. Accordingly,

$$p = P(A) = \frac{\text{area of } A}{\text{area of } S} = \frac{\pi(\frac{1}{4})^2}{1} = \frac{\pi}{16} \approx .2$$

Note. We cannot take S to be all of $\mathbf{R}^2$ because the latter has infinite area.

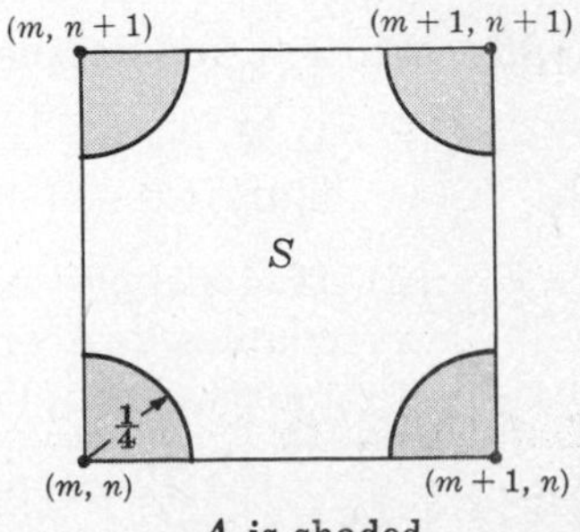

A is shaded.

3.17. Three points a, b and c are selected at random from the circumference of a circle. Find the probability p that the points lie on a semicircle.

Suppose the length of the circumference is $2s$. Let x denote the clockwise arc length from a to b, and let y denote the clockwise arc length from a to c. Thus

$$0 < x < 2s \quad \text{and} \quad 0 < y < 2s \qquad (*)$$

Let S denote the set of points in $\mathbf{R}^2$ for which condition (*) holds. Let A denote the subset of S for which any of the following conditions holds:

(i) $x, y < s$ (iii) $x < s$ and $y - x > s$

(ii) $x, y > s$ (iv) $y < s$ and $x - y > s$

Then A consists of those points for which a, b and c lie on a semicircle. Thus

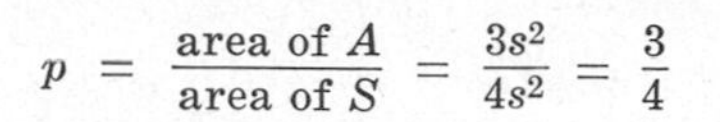

$$p = \frac{\text{area of } A}{\text{area of } S} = \frac{3s^2}{4s^2} = \frac{3}{4}$$

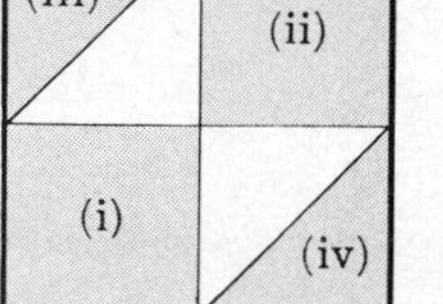

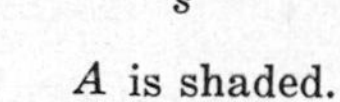

A is shaded.

MISCELLANEOUS PROBLEMS

3.18. Let A and B be events with $P(A) = \frac{3}{8}$, $P(B) = \frac{1}{2}$ and $P(A \cap B) = \frac{1}{4}$. Find (i) $P(A \cup B)$, (ii) $P(A^c)$ and $P(B^c)$, (iii) $P(A^c \cap B^c)$, (iv) $P(A^c \cup B^c)$, (v) $P(A \cap B^c)$, (vi) $P(B \cap A^c)$.

(i) $P(A \cup B) = P(A) + P(B) - P(A \cap B) = \frac{3}{8} + \frac{1}{2} - \frac{1}{4} = \frac{5}{8}$

(ii) $P(A^c) = 1 - P(A) = 1 - \frac{3}{8} = \frac{5}{8}$ and $P(B^c) = 1 - P(B) = 1 - \frac{1}{2} = \frac{1}{2}$

(iii) Using De Morgan's Law, $(A \cup B)^c = A^c \cap B^c$, we have

$$P(A^c \cap B^c) = P((A \cup B)^c) = 1 - P(A \cup B) = 1 - \tfrac{5}{8} = \tfrac{3}{8}$$

(iv) Using De Morgan's Law, $(A \cap B)^c = A^c \cup B^c$, we have

$$P(A^c \cup B^c) = P((A \cap B)^c) = 1 - P(A \cap B) = 1 - \tfrac{1}{4} = \tfrac{3}{4}$$

Equivalently,

$$P(A^c \cup B^c) = P(A^c) + P(B^c) - P(A^c \cap B^c) = \tfrac{5}{8} + \tfrac{1}{2} - \tfrac{3}{8} = \tfrac{3}{4}$$

(v) $P(A \cap B^c) = P(A \setminus B) = P(A) - P(A \cap B) = \frac{3}{8} - \frac{1}{4} = \frac{1}{8}$

(vi) $P(B \cap A^c) = P(B) - P(A \cap B) = \frac{1}{2} - \frac{1}{4} = \frac{1}{4}$

3.19. Let A and B be events with $P(A \cup B) = \frac{3}{4}$, $P(A^c) = \frac{2}{3}$ and $P(A \cap B) = \frac{1}{4}$. Find (i) $P(A)$, (ii) $P(B)$, (iii) $P(A \cap B^c)$.

(i) $P(A) = 1 - P(A^c) = 1 - \frac{2}{3} = \frac{1}{3}$

(ii) Substitute in $P(A \cup B) = P(A) + P(B) - P(A \cap B)$ to obtain $\frac{3}{4} = \frac{1}{3} + P(B) - \frac{1}{4}$ or $P(B) = \frac{2}{3}$.

(iii) $P(A \cap B^c) = P(A) - P(A \cap B) = \frac{1}{3} - \frac{1}{4} = \frac{1}{12}$

3.20. Find the probability p of an event if the odds that it will occur are $a:b$, that is, "a to b".

The odds that an event with probability p occurs is the ratio $p : (1 - p)$. Hence

$$\frac{p}{1-p} = \frac{a}{b} \quad \text{or} \quad bp = a - ap \quad \text{or} \quad ap + bp = a \quad \text{or} \quad p = \frac{a}{a+b}$$

3.21. Find the probability p of an event if the odds that it will occur are "3 to 2".

$\frac{p}{1-p} = \frac{3}{2}$ from which $p = \frac{3}{5}$. We can also use the formula of the preceding problem to obtain the answer directly: $p = \frac{a}{a+b} = \frac{3}{3+2} = \frac{3}{5}$.

3.22. A die is tossed 100 times. The following table lists the six numbers and frequency with which each number appeared:

Number	1	2	3	4	5	6
Frequency	14	17	20	18	15	16

Find the relative frequency f of the event (i) a 3 appears, (ii) a 5 appears, (iii) an even number appears, (iv) a prime appears.

The relative frequency $f = \dfrac{\text{number of successes}}{\text{total number of trials}}$.

(i) $f = \frac{20}{100} = .20$ (ii) $f = \frac{15}{100} = .15$ (iii) $f = \frac{17+18+16}{100} = .51$ (iv) $f = \frac{17+20+15}{100} = .52$

3.23. Prove Corollary 3.6: For any events A, B and C,

$$P(A \cup B \cup C) = P(A) + P(B) + P(C) - P(A \cap B) - P(A \cap C) - P(B \cap C) + P(A \cap B \cap C)$$

Let $D = B \cup C$. Then $A \cap D = A \cap (B \cup C) = (A \cap B) \cup (A \cap C)$ and

$$P(A \cap D) = P(A \cap B) + P(A \cap C) - P(A \cap B \cap A \cap C) = P(A \cap B) + P(A \cap C) - P(A \cap B \cap C)$$

Thus

$$\begin{aligned} P(A \cup B \cup C) &= P(A \cup D) = P(A) + P(D) - P(A \cap D) \\ &= P(A) + P(B) + P(C) - P(B \cap C) - [P(A \cap B) + P(A \cap C) - P(A \cap B \cap C)] \\ &= P(A) + P(B) + P(C) - P(B \cap C) - P(A \cap B) - P(A \cap C) + P(A \cap B \cap C) \end{aligned}$$

3.24. Let $S = \{a_1, a_2, \ldots, a_s\}$ and $T = \{b_1, b_2, \ldots, b_t\}$ be finite probability spaces. Let the number $p_{ij} = P(a_i)\,P(b_j)$ be assigned to the ordered pair (a_i, b_j) in the product set $S \times T = \{(s,t) : s \in S,\ t \in T\}$. Show that the p_{ij} define a probability space on $S \times T$, i.e. that the p_{ij} are nonnegative and add up to one. (This is called the *product probability space*. We emphasize that this is not the only probability function that can be defined on the product set $S \times T$.)

Since $P(a_i), P(b_j) \geqq 0$, for each i and each j, $p_{ij} = P(a_i)\,P(b_j) \geqq 0$. Furthermore,

$$\begin{aligned} & p_{11} + p_{12} + \cdots + p_{1t} + p_{21} + p_{22} + \cdots + p_{2t} + \cdots + p_{s1} + p_{s2} + \cdots + p_{st} \\ &= P(a_1)\,P(b_1) + \cdots + P(a_1)\,P(b_t) + \cdots + P(a_s)\,P(b_1) + \cdots + P(a_s)\,P(b_t) \\ &= P(a_1)[P(b_1) + \cdots + P(b_t)] + \cdots + P(a_s)[P(b_1) + \cdots + P(b_t)] \\ &= P(a_1) \cdot 1 + \cdots + P(a_s) \cdot 1 \\ &= P(a_1) + \cdots + P(a_s) \\ &= 1 \end{aligned}$$

Supplementary Problems

SAMPLE SPACES AND EVENTS

3.25. Let A and B be events. Find an expression and exhibit the Venn diagram for the event that (i) A or not B occurs, (ii) neither A nor B occurs.

3.26. Let A, B and C be events. Find an expression and exhibit the Venn diagram for the event that (i) exactly one of the three events occurs, (ii) at least two of the events occurs, (iii) none of the events occurs, (iv) A or B, but not C, occurs.

3.27. Let a penny, a dime and a die be tossed.

(i) Describe a suitable sample space S.

(ii) Express explicitly the following events: A = {two heads and an even number appear}, B = {a 2 appears}, C = {exactly one head and a prime number appear}.

(iii) Express explicitly the event that (*a*) A and B occur, (*b*) only B occurs, (*c*) B or C occurs.

FINITE PROBABILITY SPACES

3.28. Which function defines a probability space on $S = \{a_1, a_2, a_3\}$?

(i) $P(a_1) = \frac{1}{4}$, $P(a_2) = \frac{1}{3}$, $P(a_3) = \frac{1}{2}$ (iii) $P(a_1) = \frac{1}{6}$, $P(a_2) = \frac{1}{3}$, $P(a_3) = \frac{1}{2}$

(ii) $P(a_1) = \frac{2}{3}$, $P(a_2) = -\frac{1}{3}$, $P(a_3) = \frac{2}{3}$ (iv) $P(a_1) = 0$, $P(a_2) = \frac{1}{3}$, $P(a_3) = \frac{2}{3}$

3.29. Let P be a probability function on $S = \{a_1, a_2, a_3\}$. Find $P(a_1)$ if (i) $P(a_2) = \frac{1}{3}$ and $P(a_3) = \frac{1}{4}$, (ii) $P(a_1) = 2P(a_2)$ and $P(a_3) = \frac{1}{4}$, (iii) $P(\{a_2, a_3\}) = 2P(a_1)$, (iv) $P(a_3) = 2P(a_2)$ and $P(a_2) = 3P(a_1)$.

3.30. A coin is weighted so that heads is three times as likely to appear as tails. Find $P(\text{H})$ and $P(\text{T})$.

3.31. Three students A, B and C are in a swimming race. A and B have the same probability of winning and each is twice as likely to win as C. Find the probability that B or C wins.

3.32. A die is weighted so that the even numbers have the same chance of appearing, the odd numbers have the same chance of appearing, and each even number is twice as likely to appear as any odd number. Find the probability that (i) an even number appears, (ii) a prime number appears, (iii) an odd number appears, (iv) an odd prime number appears.

3.33. Find the probability of an event if the odds that it will occur are (i) 2 to 1, (ii) 5 to 11.

3.34. In a swimming race, the odds that A will win are 2 to 3 and the odds that B will win are 1 to 4. Find the probability p and the odds that A or B wins the race.

FINITE EQUIPROBABLE SPACES

3.35. A class contains 5 freshmen, 4 sophomores, 8 juniors and 3 seniors. A student is chosen at random to represent the class. Find the probability that the student is (i) a sophomore, (ii) a senior, (iii) a junior or senior.

3.36. One card is selected at random from 50 cards numbered 1 to 50. Find the probability that the number on the card is (i) divisible by 5, (ii) prime, (iii) ends in the digit 2.

3.37. Of 10 girls in a class, 3 have blue eyes. If two of the girls are chosen at random, what is the probability that (i) both have blue eyes, (ii) neither has blue eyes, (iii) at least one has blue eyes?

3.38. Three bolts and three nuts are put in a box. If two parts are chosen at random, find the probability that one is a bolt and one a nut.

3.39. Ten students, $A, B, \ldots$, are in a class. If a committee of 3 is chosen at random from the class, find the probability that (i) A belongs to the committee, (ii) B belongs to the committee, (iii) A and B belong to the committee, (iv) A or B belongs to the committee.

3.40. A class consists of 6 girls and 10 boys. If a committee of 3 is chosen at random from the class, find the probability that (i) 3 boys are selected, (ii) exactly 2 boys are selected, (iii) at least one boy is selected, (iv) exactly 2 girls are selected.

3.41. A pair of fair dice is tossed. Find the probability that the maximum of the two numbers is greater than 4.

3.42. Of 120 students, 60 are studying French, 50 are studying Spanish, and 20 are studying French and Spanish. If a student is chosen at random, find the probability that the student (i) is studying French or Spanish, (ii) is studying neither French nor Spanish.

3.43. Three boys and 3 girls sit in a row. Find the probability that (i) the 3 girls sit together, (ii) the boys and girls sit in alternate seats.

NONCOUNTABLE UNIFORM SPACES

3.44. A point is selected at random inside an equilateral triangle whose side length is 3. Find the probability that its distance to any corner is greater than 1.

3.45. A coin of diameter $\frac{1}{2}$ is tossed randomly onto the Cartesian plane $\mathbf{R}^2$. Find the probability that the coin does not intersect any line whose equation is of the form (*a*) $x = k$, (*b*) $x + y = k$, (*c*) $x = k$ or $y = k$. (Here k is an integer.)

3.46. A point X is selected at random from a line segment AB with midpoint O. Find the probability that the line segments AX, XB and AO can form a triangle.

MISCELLANEOUS PROBLEMS

3.47. Let A and B be events with $P(A \cup B) = \frac{7}{8}$, $P(A \cap B) = \frac{1}{4}$ and $P(A^c) = \frac{5}{8}$. Find $P(A)$, $P(B)$ and $P(A \cap B^c)$.

3.48. Let A and B be events with $P(A) = \frac{1}{2}$, $P(A \cup B) = \frac{3}{4}$ and $P(B^c) = \frac{5}{8}$. Find $P(A \cap B)$, $P(A^c \cap B^c)$, $P(A^c \cup B^c)$ and $P(B \cap A^c)$.

3.49. A die is tossed 50 times. The following table gives the six numbers and their frequency of occurrence:

Number	1	2	3	4	5	6
Frequency	7	9	8	7	9	10

Find the relative frequency of the event (i) a 4 appears, (ii) an odd number appears, (iii) a prime number appears.

3.50. Prove: For any events $A_1, A_2, \ldots, A_n$,

$$P(A_1 \cup \cdots \cup A_n) = \sum_i P(A_i) - \sum_{i<j} P(A_i \cap A_j) + \sum_{i<j<k} P(A_i \cap A_j \cap A_k) - \cdots \pm P(A_1 \cap \cdots \cap A_n)$$

(*Remark*: This result generalizes Theorem 3.5 and Corollary 3.6.)

Answers to Supplementary Problems

3.25. (i) $A \cup B^c$, (ii) $(A \cup B)^c$

3.26. (i) $(A \cap B^c \cap C^c) \cup (B \cap A^c \cap C^c) \cup (C \cap A^c \cap B^c)$ (iii) $(A \cup B \cup C)^c$
(ii) $(A \cap B) \cup (A \cap C) \cup (B \cap C)$ (iv) $(A \cup B) \cap C^c$

3.27. (i) S = {HH1, HH2, HH3, HH4, HH5, HH6, HT1, HT2, HT3, HT4, HT5, HT6, TH1, TH2, TH3, TH4, TH5, TH6, TT1, TT2, TT3, TT4, TT5, TT6}
(ii) A = {HH2, HH4, HH6}, B = {HH2, HT2, TH2, TT2}, C = {HT2, TH2, HT3, TH3, HT5, TH5}
(iii) (a) $A \cap B$ = {HH2}
(b) $B \setminus (A \cup C)$ = {TT2}
(c) $B \cup C$ = {HH2, HT2, TH2, TT2, HT3, TH3, HT5, TH5}

3.28. (i) no, (ii) no, (iii) yes, (iv) yes

3.29. (i) $\frac{5}{12}$, (ii) $\frac{1}{2}$, (iii) $\frac{1}{3}$, (iv) $\frac{1}{10}$

3.30. $P(H) = \frac{3}{4}$, $P(T) = \frac{1}{4}$

3.31. $\frac{3}{5}$

3.32. (i) $\frac{2}{3}$, (ii) $\frac{4}{9}$, (iii) $\frac{1}{3}$, (v) $\frac{2}{9}$

3.33. (i) $\frac{2}{3}$, (ii) $\frac{5}{16}$

3.34. $p = \frac{3}{5}$; the odds are 3 to 2.

3.35. (i) $\frac{1}{5}$, (ii) $\frac{3}{20}$, (iii) $\frac{11}{20}$

3.36. (i) $\frac{1}{5}$, (ii) $\frac{3}{10}$, (iii) $\frac{1}{10}$

3.37. (i) $\frac{1}{15}$, (ii) $\frac{7}{15}$, (iii) $\frac{8}{15}$

3.38. $\frac{3}{5}$

3.39. (i) $\frac{3}{10}$, (ii) $\frac{3}{10}$, (iii) $\frac{1}{15}$, (iv) $\frac{8}{15}$

3.40. (i) $\frac{3}{14}$, (ii) $\frac{27}{56}$, (iii) $\frac{27}{28}$, (iv) $\frac{15}{56}$

3.41. $\frac{5}{9}$

3.42. (i) $\frac{3}{4}$, (ii) $\frac{1}{4}$

3.43. (i) $\frac{1}{5}$, (ii) $\frac{1}{10}$

3.44. $1 - 2\pi/(9\sqrt{3})$

3.45. (i) $\frac{1}{2}$, (ii) $1 - \frac{1}{2}\sqrt{2}$, (iii) $\frac{1}{4}$

3.46. $\frac{1}{2}$

3.47. $P(A) = \frac{3}{8}$, $P(B) = \frac{3}{4}$, $P(A \cap B^c) = \frac{1}{8}$

3.48. $P(A \cap B) = \frac{1}{8}$, $P(A^c \cap B^c) = \frac{1}{4}$, $P(A^c \cup B^c) = \frac{7}{8}$, $P(B \cap A^c) = \frac{1}{4}$

3.49. (i) $\frac{7}{50}$, (ii) $\frac{24}{50}$, (iii) $\frac{26}{50}$

Chapter 4

Conditional Probability and Independence

CONDITIONAL PROBABILITY

Let E be an arbitrary event in a sample space S with $P(E) > 0$. The probability that an event A occurs once E has occurred or, in other words, the *conditional probability* of A given E, written $P(A \mid E)$, is defined as follows:

$$P(A \mid E) = \frac{P(A \cap E)}{P(E)}$$

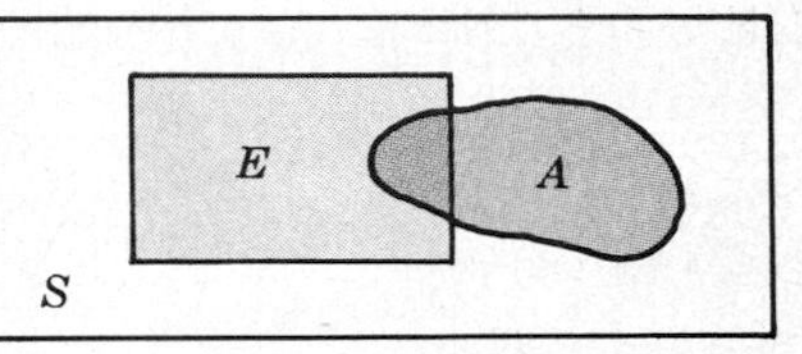

As seen in the adjoining Venn diagram, $P(A \mid E)$ in a certain sense measures the relative probability of A with respect to the reduced space E.

In particular, if S is a finite equiprobable space and $|A|$ denotes the number of elements in an event A, then

$$P(A \cap E) = \frac{|A \cap E|}{|S|}, \quad P(E) = \frac{|E|}{|S|} \quad \text{and so} \quad P(A \mid E) = \frac{P(A \cap E)}{P(E)} = \frac{|A \cap E|}{|E|}$$

That is,

Theorem 4.1: Let S be a finite equiprobable space with events A and E. Then

$$P(A \mid E) = \frac{\text{number of elements in } A \cap E}{\text{number of elements in } E}$$

or

$$P(A \mid E) = \frac{\text{number of ways } A \text{ and } E \text{ can occur}}{\text{number of ways } E \text{ can occur}}$$

Example 4.1: Let a pair of fair dice be tossed. If the sum is 6, find the probability that one of the dice is a 2. In other words, if

$$E = \{\text{sum is 6}\} = \{(1,5), (2,4), (3,3), (4,2), (5,1)\}$$

and $$A = \{\text{a 2 appears on at least one die}\}$$

find $P(A \mid E)$.

Now E consists of five elements and two of them, $(2,4)$ and $(4,2)$, belong to A: $A \cap E = \{(2,4), (4,2)\}$. Then $P(A \mid E) = \frac{2}{5}$.

On the other hand, since A consists of eleven elements,

$$A = \{(2,1), (2,2), (2,3), (2,4), (2,5), (2,6), (1,2), (3,2), (4,2), (5,2), (6,2)\}$$

and S consists of 36 elements, $P(A) = \frac{11}{36}$.

Example 4.2: A man visits a couple who have two children. One of the children, a boy, comes into the room. Find the probability p that the other is also a boy if (i) the other child is known to be younger, (ii) nothing is known about the other child.

The sample space for the sex of two children is $S = \{bb, bg, gb, gg\}$ with probability $\frac{1}{4}$ for each point. (Here the sequence of each point corresponds to the sequence of births.)

(i) The reduced sample space consists of two elements, $\{bb, bg\}$; hence $p = \frac{1}{2}$.

(ii) The reduced sample space consists of three elements, $\{bb, bg, gb\}$; hence $p = \frac{1}{3}$.

MULTIPLICATION THEOREM FOR CONDITIONAL PROBABILITY

If we cross multiply the above equation defining conditional probability and use the fact that $A \cap E = E \cap A$, we obtain the following useful formula.

Theorem 4.2: $P(E \cap A) = P(E)\,P(A \mid E)$

This theorem can be extended by induction as follows:

Corollary 4.3: For any events $A_1, A_2, \ldots, A_n$,

$$P(A_1 \cap A_2 \cap \cdots \cap A_n)$$
$$= P(A_1)\,P(A_2 \mid A_1)\,P(A_3 \mid A_1 \cap A_2) \cdots P(A_n \mid A_1 \cap A_2 \cap \cdots \cap A_{n-1})$$

We now apply the above theorem which is called, appropriately, the *multiplication theorem.*

Example 4.3: A lot contains 12 items of which 4 are defective. Three items are drawn at random from the lot one after the other. Find the probability p that all three are nondefective.

The probability that the first item is nondefective is $\frac{8}{12}$ since 8 of 12 items are nondefective. If the first item is nondefective, then the probability that the next item is nondefective is $\frac{7}{11}$ since only 7 of the remaining 11 items are nondefective. If the first two items are nondefective, then the probability that the last item is nondefective is $\frac{6}{10}$ since only 6 of the remaining 10 items are now nondefective. Thus by the multiplication theorem,

$$p = \frac{8}{12} \cdot \frac{7}{11} \cdot \frac{6}{10} = \frac{14}{55}$$

FINITE STOCHASTIC PROCESSES AND TREE DIAGRAMS

A (finite) sequence of experiments in which each experiment has a finite number of outcomes with given probabilities is called a *(finite) stochastic process.* A convenient way of describing such a process and computing the probability of any event is by a *tree diagram* as illustrated below; the multiplication theorem of the previous section is used to compute the probability that the result represented by any given path of the tree does occur.

Example 4.4: We are given three boxes as follows:

Box I has 10 light bulbs of which 4 are defective.

Box II has 6 light bulbs of which 1 is defective.

Box III has 8 light bulbs of which 3 are defective.

We select a box at random and then draw a bulb at random. What is the probability p that the bulb is defective?

Here we perform a sequence of two experiments:

(i) select one of the three boxes;

(ii) select a bulb which is either defective (D) or nondefective (N).

The following tree diagram describes this process and gives the probability of each branch of the tree:

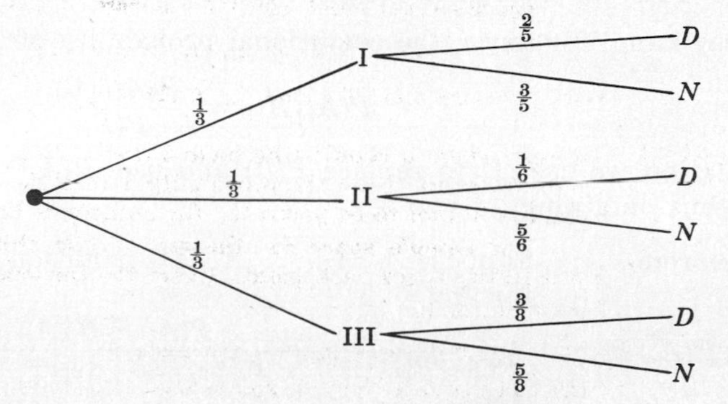

The probability that any particular path of the tree occurs is, by the multiplication theorem, the product of the probabilities of each branch of the path, e.g., the probability of selecting box I and then a defective bulb is $\frac{1}{3} \cdot \frac{2}{5} = \frac{2}{15}$.

Now since there are three mutually exclusive paths which lead to a defective bulb, the sum of the probabilities of these paths is the required probability:

$$p = \frac{1}{3} \cdot \frac{2}{5} + \frac{1}{3} \cdot \frac{1}{6} + \frac{1}{3} \cdot \frac{3}{8} = \frac{113}{360}$$

Example 4.5: A coin, weighted so that $P(\text{H}) = \frac{2}{3}$ and $P(\text{T}) = \frac{1}{3}$, is tossed. If heads appears, then a number is selected at random from the numbers 1 through 9; if tails appears, then a number is selected at random from the numbers 1 through 5. Find the probability p that an even number is selected.

The tree diagram with respective probabilities is

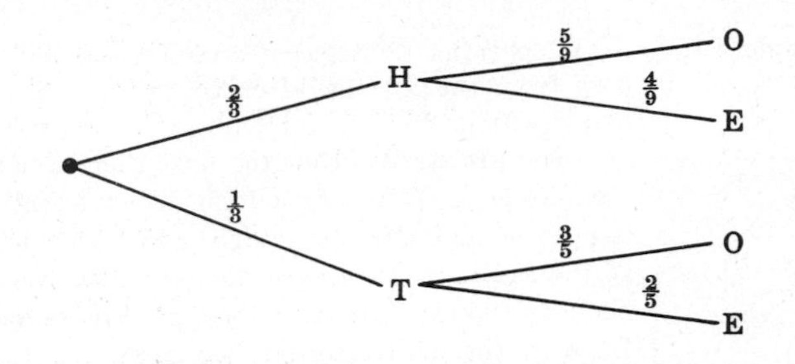

Note that the probability of selecting an even number from the numbers 1 through 9 is $\frac{4}{9}$ since there are 4 even numbers out of the 9 numbers, whereas the probability of selecting an even number from the numbers 1 through 5 is $\frac{2}{5}$ since there are 2 even numbers out of the 5 numbers. Two of the paths lead to an even number: HE and TE. Thus

$$p = P(\text{E}) = \frac{2}{3} \cdot \frac{4}{9} + \frac{1}{3} \cdot \frac{2}{5} = \frac{58}{135}$$

PARTITIONS AND BAYES' THEOREM

Suppose the events $A_1, A_2, \ldots, A_n$ form a partition of a sample space S; that is, the events A_i are mutually exclusive and their union is S. Now let B be any other event. Then

$$B = S \cap B = (A_1 \cup A_2 \cup \cdots \cup A_n) \cap B$$
$$= (A_1 \cap B) \cup (A_2 \cap B) \cup \cdots \cup (A_n \cap B)$$

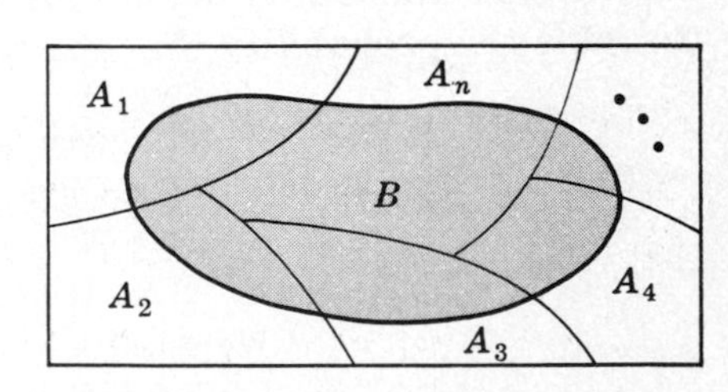

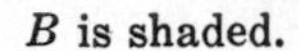
B is shaded.

where the $A_i \cap B$ are also mutually exclusive. Accordingly,

$$P(B) = P(A_1 \cap B) + P(A_2 \cap B) + \cdots + P(A_n \cap B)$$

Thus by the multiplication theorem,

$$P(B) = P(A_1)\,P(B \mid A_1) + P(A_2)\,P(B \mid A_2) + \cdots + P(A_n)\,P(B \mid A_n) \tag{1}$$

On the other hand, for any i, the conditional probability of A_i given B is defined by

$$P(A_i \mid B) = \frac{P(A_i \cap B)}{P(B)}$$

In this equation we use (*1*) to replace $P(B)$ and use $P(A_i \cap B) = P(A_i)\,P(B \mid A_i)$ to replace $P(A_i \cap B)$, thus obtaining

Bayes' Theorem 4.4: Suppose $A_1, A_2, \ldots, A_n$ is a partition of S and B is any event. Then for any i,

$$P(A_i \mid B) = \frac{P(A_i)\,P(B \mid A_i)}{P(A_1)\,P(B \mid A_1) + P(A_2)\,P(B \mid A_2) + \cdots + P(A_n)\,P(B \mid A_n)}$$

Example 4.6: Three machines A, B and C produce respectively 50%, 30% and 20% of the total number of items of a factory. The percentages of defective output of these machines are 3%, 4% and 5%. If an item is selected at random, find the probability that the item is defective.

Let X be the event that an item is defective. Then by (*1*) above,

$$\begin{aligned} P(X) &= P(A)\,P(X \mid A) + P(B)\,P(X \mid B) \\ &\quad + P(C)\,P(X \mid C) \\ &= (.50)(.03) + (.30)(.04) + (.20)(.05) \\ &= .037 \end{aligned}$$

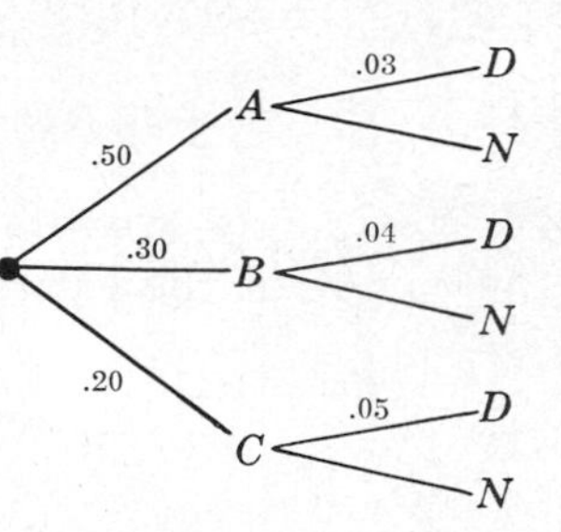

Observe that we can also consider this problem as a stochastic process having the adjoining tree diagram.

Example 4.7: Consider the factory in the preceding example. Suppose an item is selected at random and is found to be defective. Find the probability that the item was produced by machine A; that is, find $P(A \mid X)$.

By Bayes' theorem,

$$\begin{aligned} P(A \mid X) &= \frac{P(A)\,P(X \mid A)}{P(A)\,P(X \mid A) + P(B)\,P(X \mid B) + P(C)\,P(X \mid C)} \\ &= \frac{(.50)(.03)}{(.50)(.03) + (.30)(.04) + (.20)(.05)} = \frac{15}{37} \end{aligned}$$

In other words, we divide the probability of the required path by the probability of the reduced sample space, i.e. those paths which lead to a defective item.

INDEPENDENCE

An event B is said to be *independent* of an event A if the probability that B occurs is not influenced by whether A has or has not occurred. In other words, if the probability of B equals the conditional probility of B given A: $P(B) = P(B \mid A)$. Now substituting $P(B)$ for $P(B \mid A)$ in the multiplications theorem $P(A \cap B) = P(A)\,P(B \mid A)$, we obtain

$$P(A \cap B) = P(A)\,P(B)$$

We use the above equation as our formal definition of independence.

Definition: Events A and B are independent if $P(A \cap B) = P(A)\,P(B)$; otherwise they are dependent.

Example 4.8: Let a fair coin be tossed three times; we obtain the equiprobable space

$$S = \{\text{HHH, HHT, HTH, HTT, THH, THT, TTH, TTT}\}$$

Consider the events

$$A = \{\text{first toss is heads}\}, \qquad B = \{\text{second toss is heads}\}$$
$$C = \{\text{exactly two heads are tossed in a row}\}$$

Clearly A and B are independent events; this fact is verified below. On the other hand, the relationship between A and C or B and C is not obvious. We claim that A and C are independent, but that B and C are dependent. We have

$$P(A) = P(\{\text{HHH, HHT, HTH, HTT}\}) = \tfrac{4}{8} = \tfrac{1}{2}$$

$$P(B) = P(\{\text{HHH, HHT, THH, THT}\}) = \tfrac{4}{8} = \tfrac{1}{2}$$

$$P(C) = P(\{\text{HHT, THH}\}) = \tfrac{2}{8} = \tfrac{1}{4}$$

Then

$$P(A \cap B) = P(\{\text{HHH, HHT}\}) = \tfrac{1}{4}, \qquad P(A \cap C) = P(\{\text{HHT}\}) = \tfrac{1}{8},$$
$$P(B \cap C) = P(\{\text{HHT, THH}\}) = \tfrac{1}{4}$$

Accordingly,

$$P(A)\,P(B) = \frac{1}{2}\cdot\frac{1}{2} = \frac{1}{4} = P(A\cap B), \quad \text{and so } A \text{ and } B \text{ are independent;}$$

$$P(A)\,P(C) = \frac{1}{2}\cdot\frac{1}{4} = \frac{1}{8} = P(A\cap C), \quad \text{and so } A \text{ and } C \text{ are independent;}$$

$$P(B)\,P(C) = \frac{1}{2}\cdot\frac{1}{4} = \frac{1}{8} \neq P(B\cap C), \quad \text{and so } B \text{ and } C \text{ are dependent.}$$

Frequently, we will postulate that two events are independent, or it will be clear from the nature of the experiment that two events are independent.

Example 4.9: The probability that A hits a target is $\frac{1}{4}$ and the probability that B hits it is $\frac{2}{5}$. What is the probability that the target will be hit if A and B each shoot at the target?

We are given that $P(A) = \frac{1}{4}$ and $P(B) = \frac{2}{5}$, and we seek $P(A\cup B)$. Furthermore, the probability that A or B hits the target is not influenced by what the other does; that is, the event that A hits the target is independent of the event that B hits the target: $P(A\cap B) = P(A)\,P(B)$. Thus

$$P(A\cup B) = P(A) + P(B) - P(A\cap B) = P(A) + P(B) - P(A)\,P(B)$$
$$= \frac{1}{4} + \frac{2}{5} - \frac{1}{4}\cdot\frac{2}{5} = \frac{11}{20}$$

Three events A, B and C are *independent* if:

(i) $P(A\cap B) = P(A)\,P(B)$, $P(A\cap C) = P(A)\,P(C)$ and $P(B\cap C) = P(B)\,P(C)$

i.e. if the events are pairwise independent, and

(ii) $P(A\cap B\cap C) = P(A)\,P(B)\,P(C)$.

The next example shows that condition (ii) does not follow from condition (i); in other words, three events may be pairwise independent but not independent themselves.

Example 4.10: Let a pair of fair coins be tossed; here $S = \{\text{HH, HT, TH, TT}\}$ is an equiprobable space. Consider the events

$$A = \{\text{heads on the first coin}\} = \{\text{HH, HT}\}$$
$$B = \{\text{heads on the second coin}\} = \{\text{HH, TH}\}$$
$$C = \{\text{heads on exactly one coin}\} = \{\text{HT, TH}\}$$

Then $P(A) = P(B) = P(C) = \frac{2}{4} = \frac{1}{2}$ and

$$P(A\cap B) = P(\{\text{HH}\}) = \frac{1}{4}, \quad P(A\cap C) = P(\{\text{HT}\}) = \frac{1}{4}, \quad P(B\cap C) = (\{\text{TH}\}) = \frac{1}{4}$$

Thus condition (i) is satisfied, i.e., the events are pairwise independent. However, $A\cap B\cap C = \emptyset$ and so

$$P(A\cap B\cap C) = P(\emptyset) = 0 \neq P(A)\,P(B)\,P(C)$$

In other words, condition (ii) is not satisfied and so the three events are not independent.

INDEPENDENT OR REPEATED TRIALS

We have previously discussed probability spaces which were associated with an experiment repeated a finite number of times, as the tossing of a coin three times. This concept of repetition is formalized as follows:

Definition: Let S be a finite probability space. By *n independent* or *repeated trials,* we mean the probability space T consisting of ordered n-tuples of elements of S with the probability of an n-tuple defined to be the product of the probabilities of its components:

$$P((s_1, s_2, \ldots, s_n)) = P(s_1)\,P(s_2)\cdots P(s_n)$$

Example 4.11: Whenever three horses a, b and c race together, their respective probabilities of winning are $\frac{1}{2}$, $\frac{1}{3}$ and $\frac{1}{6}$. In other words, $S = \{a, b, c\}$ with $P(a) = \frac{1}{2}$, $P(b) = \frac{1}{3}$ and $P(c) = \frac{1}{6}$. If the horses race twice, then the sample space of the 2 repeated trials is

$$T = \{aa, ab, ac, ba, bb, bc, ca, cb, cc\}$$

For notational convenience, we have written ac for the ordered pair (a, c). The probability of each point in T is

$$P(aa) = P(a)\,P(a) = \frac{1}{2}\cdot\frac{1}{2} = \frac{1}{4} \qquad P(ba) = \frac{1}{6} \qquad P(ca) = \frac{1}{12}$$

$$P(ab) = P(a)\,P(b) = \frac{1}{2}\cdot\frac{1}{3} = \frac{1}{6} \qquad P(bb) = \frac{1}{9} \qquad P(cb) = \frac{1}{18}$$

$$P(ac) = P(a)\,P(c) = \frac{1}{2}\cdot\frac{1}{6} = \frac{1}{12} \qquad P(bc) = \frac{1}{18} \qquad P(cc) = \frac{1}{36}$$

Thus the probability of c winning the first race and a winning the second race is $P(ca) = \frac{1}{12}$.

From another point of view, a repeated trials process is a stochastic process whose tree diagram has the following properties: (i) every branch point has the same outcomes; (ii) the probability is the same for each branch leading to the same outcome. For example, the tree diagram of the repeated trials process of the preceding experiment is as shown in the adjoining figure.

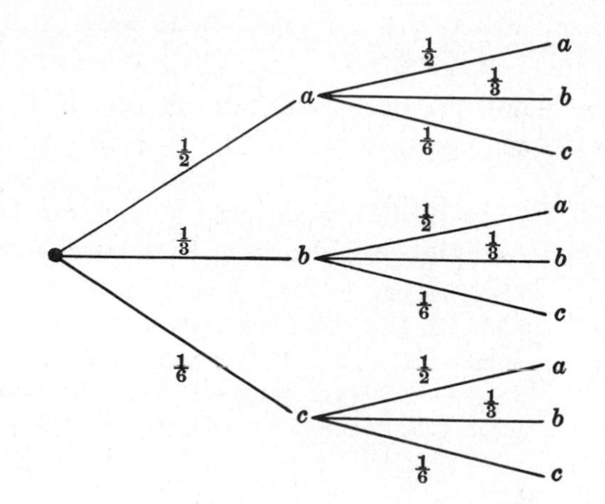

Observe that every branch point has the outcomes a, b and c, and each branch leading to outcome a has probability $\frac{1}{2}$, each branch leading to b has probability $\frac{1}{3}$, and each leading to c has probability $\frac{1}{6}$.

Solved Problems

CONDITIONAL PROBABILITY IN FINITE EQUIPROBABLE SPACES

4.1. A pair of fair dice is thrown. Find the probability p that the sum is 10 or greater if (i) a 5 appears on the first die, (ii) a 5 appears on at least one of the dice.

(i) If a 5 appears on the first die, then the reduced sample space is

$$A = \{(5,1), (5,2), (5,3), (5,4), (5,5), (5,6)\}$$

The sum is 10 or greater on two of the six outcomes: $(5,5)$, $(5,6)$. Hence $p = \frac{2}{6} = \frac{1}{3}$.

(ii) If a 5 appears on at least one of the dice, then the reduced sample space has eleven elements:

$$B = \{(5,1), (5,2), (5,3), (5,4), (5,5), (5,6), (1,5), (2,5), (3,5), (4,5), (6,5)\}$$

The sum is 10 or greater on three of the eleven outcomes: $(5,5)$, $(5,6)$, $(6,5)$. Hence $p = \frac{3}{11}$.

4.2. Three fair coins are tossed. Find the probability p that they are all heads if (i) the first coin is heads, (ii) one of the coins is heads.

The sample space has eight elements: $S = \{HHH, HHT, HTH, HTT, THH, THT, TTH, TTT\}$.

(i) If the first coin is heads, the reduced sample space is $A = \{HHH, HHT, HTH, HTT\}$. Since the coins are all heads in 1 of 4 cases, $p = \frac{1}{4}$.

(ii) If one of the coins is heads, the reduced sample space is $B = \{HHH, HHT, HTH, HTT, THH, THT, TTH\}$. Since the coins are all heads in 1 of 7 cases, $p = \frac{1}{7}$.

4.3. A pair of fair dice is thrown. If the two numbers appearing are different, find the probability p that (i) the sum is six, (ii) an ace appears, (iii) the sum is 4 or less.

Of the 36 ways the pair of dice can be thrown, 6 will contain the same numbers: $(1,1)$, $(2,2)$, $\ldots$, $(6,6)$. Thus the reduced sample space will consist of $36 - 6 = 30$ elements.

(i) The sum 6 can appear in 4 ways: $(1,5)$, $(2,4)$, $(4,2)$, $(5,1)$. (We cannot include $(3,3)$ since the numbers are the same.) Hence $p = \frac{4}{30} = \frac{2}{15}$.

(ii) An ace can appear in 10 ways: $(1,2)$, $(1,3)$, $\ldots$, $(1,6)$ and $(2,1)$, $(3,1)$, $\ldots$, $(6,1)$. Hence $p = \frac{10}{30} = \frac{1}{3}$.

(iii) The sum of 4 or less can occur in 4 ways: $(3,1)$, $(1,3)$, $(2,1)$, $(1,2)$. Thus $p = \frac{4}{30} = \frac{2}{15}$.

4.4. Two digits are selected at random from the digits 1 through 9. If the sum is even, find the probability p that both numbers are odd.

The sum is even if both numbers are even or if both numbers are odd. There are 4 even numbers $(2,4,6,8)$; hence there are $\binom{4}{2} = 6$ ways to choose two even numbers. There are 5 odd numbers $(1,3,5,7,9)$; hence there are $\binom{5}{2} = 10$ ways to choose two odd numbers. Thus there are $6 + 10 = 16$ ways to choose two numbers such that their sum is even; since 10 of these ways occur when both numbers are odd, $p = \frac{10}{16} = \frac{5}{8}$.

4.5. A man is dealt 4 spade cards from an ordinary deck of 52 cards. If he is given three more cards, find the probability p that at least one of the additional cards is also a spade.

Since he is dealt 4 spades, there are $52 - 4 = 48$ cards remaining of which $13 - 4 = 9$ are spades. There are $\binom{48}{3} = 17{,}296$ ways in which he can be dealt three more cards. Since there are $48 - 9 = 39$ cards which are not spades, there are $\binom{39}{3} = 9139$ ways he can be dealt three cards which are not spades. Thus the probability q that he is not dealt another spade is $q = \frac{9139}{17{,}296}$; hence $p = 1 - q = \frac{8157}{17{,}296}$.

4.6. Four people, called North, South, East and West, are each dealt 13 cards from an ordinary deck of 52 cards.

(i) If South has no aces, find the probability p that his partner North has exactly two aces.

(ii) If North and South together have nine hearts, find the probability p that East and West each has two hearts.

(i) There are 39 cards, including 4 aces, divided among North, East and West. There are $\binom{39}{13}$ ways that North can be dealt 13 of the 39 cards. There are $\binom{4}{2}$ ways he can be dealt 2 of the four aces, and $\binom{35}{11}$ ways he can be dealt 11 cards from the $39 - 4 = 35$ cards which are not aces. Thus

$$p = \frac{\binom{4}{2}\binom{35}{11}}{\binom{39}{13}} = \frac{6 \cdot 12 \cdot 13 \cdot 25 \cdot 26}{36 \cdot 37 \cdot 38 \cdot 39} = \frac{650}{2109}$$

(ii) There are 26 cards, including 4 hearts, divided among East and West. There are $\binom{26}{13}$ ways that, say, East can be dealt 13 cards. (We need only analyze East's 13 cards since West must have the remaining cards.) There are $\binom{4}{2}$ ways East can be dealt 2 hearts from 4 hearts, and $\binom{22}{11}$ ways he can be dealt 11 non-hearts from the $26 - 4 = 22$ non-hearts. Thus

$$p = \frac{\binom{4}{2}\binom{22}{11}}{\binom{26}{13}} = \frac{6 \cdot 12 \cdot 13 \cdot 12 \cdot 13}{23 \cdot 24 \cdot 25 \cdot 26} = \frac{234}{575}$$

MULTIPLICATION THEOREM

4.7. A class has 12 boys and 4 girls. If three students are selected at random from the class, what is the probability p that they are all boys?

The probability that the first student selected is a boy is 12/16 since there are 12 boys out of 16 students. If the first student is a boy, then the probability that the second is a boy is 11/15 since there are 11 boys left out of 15 students. Finally, if the first two students selected were boys, then the probability that the third student is a boy is 10/14 since there are 10 boys left out of 14 students. Thus, by the multiplication theorem, the probability that all three are boys is

$$p = \frac{12}{16} \cdot \frac{11}{15} \cdot \frac{10}{14} = \frac{11}{28}$$

Another Method. There are $\binom{16}{3} = 560$ ways to select 3 students of the 16 students, and $\binom{12}{3} = 220$ ways to select 3 boys out of 12 boys; hence $p = \frac{220}{560} = \frac{11}{28}$.

A Third Method. If the students are selected one after the other, then there are $16 \cdot 15 \cdot 14$ ways to select three students, and $12 \cdot 11 \cdot 10$ ways to select three boys; hence $p = \frac{12 \cdot 11 \cdot 10}{16 \cdot 15 \cdot 14} = \frac{11}{28}$.

4.8. A man is dealt 5 cards one after the other from an ordinary deck of 52 cards. What is the probability p that they are all spades?

The probability that the first card is a spade is 13/52, the second is a spade is 12/51, the third is a spade is 11/50, the fourth is a spade is 10/49, and the last is a spade is 9/48. (We assumed in each case that the previous cards were spades.) Thus $p = \frac{13}{52} \cdot \frac{12}{51} \cdot \frac{11}{50} \cdot \frac{10}{49} \cdot \frac{9}{48} = \frac{33}{66{,}640}$.

4.9. An urn contains 7 red marbles and 3 white marbles. Three marbles are drawn from the urn one after the other. Find the probability p that the first two are red and the third is white.

The probability that the first marble is red is 7/10 since there are 7 red marbles out of 10 marbles. If the first marble is red, then the probability that the second marble is red is 6/9 since there are 6 red marbles remaining out of the 9 marbles. If the first two marbles are red, then the probability that the third marble is white is 3/8 since there are 3 white marbles out of the 8 marbles in the urn. Hence by the multiplication theorem,

$$p = \frac{7}{10} \cdot \frac{6}{9} \cdot \frac{3}{8} = \frac{7}{40}$$

4.10. The students in a class are selected at random, one after the other, for an examination. Find the probability p that the boys and girls in the class alternate if (i) the class consists of 4 boys and 3 girls, (ii) the class consists of 3 boys and 3 girls.

(i) If the boys and girls are to alternate, then the first student examined must be a boy. The probability that the first is a boy is 4/7. If the first is a boy, then the probability that the second is a girl is 3/6 since there are 3 girls out of 6 students left. Continuing in this manner, we obtain the probability that the third is a boy is 3/5, the fourth is a girl is 2/4, the fifth is a boy is 2/3, the sixth is a girl is 1/2, and the last is a boy is 1/1. Thus

$$p = \frac{4}{7} \cdot \frac{3}{6} \cdot \frac{3}{5} \cdot \frac{2}{4} \cdot \frac{2}{3} \cdot \frac{1}{2} \cdot \frac{1}{1} = \frac{1}{35}$$

(ii) There are two mutually exclusive cases: the first pupil is a boy, and the first is a girl. If the first student is a boy, then by the multiplication theorem the probability p_1 that the students alternate is

$$p_1 = \frac{3}{6}\cdot\frac{3}{5}\cdot\frac{2}{4}\cdot\frac{2}{3}\cdot\frac{1}{2}\cdot\frac{1}{1} = \frac{1}{20}$$

If the first student is a girl, then by the multiplication theorem the probability p_2 that the students alternate is

$$p_2 = \frac{3}{6}\cdot\frac{3}{5}\cdot\frac{2}{4}\cdot\frac{2}{3}\cdot\frac{1}{2}\cdot\frac{1}{1} = \frac{1}{20}$$

Thus $p = p_1 + p_2 = \frac{1}{20} + \frac{1}{20} = \frac{1}{10}$.

MISCELLANEOUS PROBLEMS ON CONDITIONAL PROBABILITY

4.11. In a certain college, 25% of the students failed mathematics, 15% of the students failed chemistry, and 10% of the students failed both mathematics and chemistry. A student is selected at random.

(i) If he failed chemistry, what is the probability that he failed mathematics?
(ii) If he failed mathematics, what is the probability that he failed chemistry?
(iii) What is the probability that he failed mathematics or chemistry?

Let $M = \{\text{students who failed mathematics}\}$ and $C = \{\text{students who failed chemistry}\}$; then

$$P(M) = .25, \quad P(C) = .15, \quad P(M \cap C) = .10$$

(i) The probability that a student failed mathematics, given that he has failed chemistry is

$$P(M \mid C) = \frac{P(M\cap C)}{P(C)} = \frac{.10}{.15} = \frac{2}{3}$$

(ii) The probability that a student failed chemistry, given that he has failed mathematics is

$$P(C \mid M) = \frac{P(C\cap M)}{P(M)} = \frac{.10}{.25} = \frac{2}{5}$$

(iii)
$$P(M\cup C) = P(M) + P(C) - P(M\cap C) = .25 + .15 - .10 = .30 = \frac{3}{10}$$

4.12. Let A and B be events with $P(A) = \frac{1}{2}$, $P(B) = \frac{1}{3}$ and $P(A\cap B) = \frac{1}{4}$. Find (i) $P(A \mid B)$, (ii) $P(B \mid A)$, (iii) $P(A \cup B)$, (iv) $P(A^c \mid B^c)$, (v) $P(B^c \mid A^c)$.

(i) $P(A \mid B) = \dfrac{P(A\cap B)}{P(B)} = \dfrac{\frac{1}{4}}{\frac{1}{3}} = \dfrac{3}{4}$ (ii) $P(B \mid A) = \dfrac{P(B\cap A)}{P(A)} = \dfrac{\frac{1}{4}}{\frac{1}{2}} = \dfrac{1}{2}$

(iii) $P(A\cup B) = P(A) + P(B) - P(A\cap B) = \dfrac{1}{2} + \dfrac{1}{3} - \dfrac{1}{4} = \dfrac{7}{12}$

(iv) First compute $P(B^c)$ and $P(A^c\cap B^c)$. $P(B^c) = 1 - P(B) = 1 - \frac{1}{3} = \frac{2}{3}$. By De Morgan's law, $(A\cup B)^c = A^c \cap B^c$; hence $P(A^c\cap B^c) = P((A\cup B)^c) = 1 - P(A\cup B) = 1 - \frac{7}{12} = \frac{5}{12}$.

Thus $P(A^c \mid B^c) = \dfrac{P(A^c\cap B^c)}{P(B^c)} = \dfrac{\frac{5}{12}}{\frac{2}{3}} = \dfrac{5}{8}$.

(v) $P(A^c) = 1 - P(A) = 1 - \frac{1}{2} = \frac{1}{2}$. Then $P(B^c \mid A^c) = \dfrac{P(B^c\cap A^c)}{P(A^c)} = \dfrac{\frac{5}{12}}{\frac{1}{2}} = \dfrac{5}{6}$.

4.13. Let A and B be events with $P(A) = \frac{3}{8}$, $P(B) = \frac{5}{8}$ and $P(A\cup B) = \frac{3}{4}$. Find $P(A \mid B)$ and $P(B \mid A)$.

First compute $P(A\cap B)$ using the formula $P(A\cup B) = P(A) + P(B) - P(A\cap B)$:

$$\frac{3}{4} = \frac{3}{8} + \frac{5}{8} - P(A\cap B) \quad \text{or} \quad P(A\cap B) = \frac{1}{4}$$

Then $P(A \mid B) = \dfrac{P(A\cap B)}{P(B)} = \dfrac{\frac{1}{4}}{\frac{5}{8}} = \dfrac{2}{5}$ and $P(B \mid A) = \dfrac{P(B\cap A)}{P(A)} = \dfrac{\frac{1}{4}}{\frac{3}{8}} = \dfrac{2}{3}$.

4.14. Find $P(B \mid A)$ if (i) A is a subset of B, (ii) A and B are mutually exclusive.

(i) If A is a subset of B, then whenever A occurs B must occur; hence $P(B \mid A) = 1$. Alternately, if A is a subset of B then $A \cap B = A$; hence

$$P(B \mid A) = \frac{P(A \cap B)}{P(A)} = \frac{P(A)}{P(A)} = 1$$

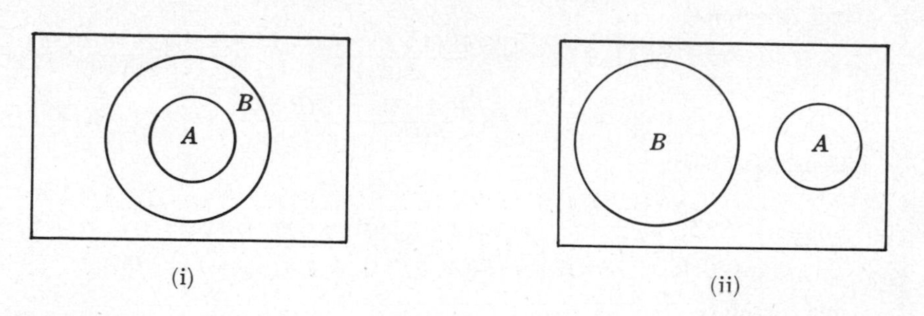

(i) (ii)

(ii) If A and B are mutually exclusive, i.e. disjoint, then whenever A occurs B cannot occur; hence $P(B \mid A) = 0$. Alternately, if A and B are mutually exclusive then $A \cap B = \emptyset$; hence

$$P(B \mid A) = \frac{P(A \cap B)}{P(A)} = \frac{P(\emptyset)}{P(A)} = \frac{0}{P(A)} = 0$$

4.15. Three machines A, B and C produce respectively 60%, 30% and 10% of the total number of items of a factory. The percentages of defective output of these machines are respectively 2%, 3% and 4%. An item is selected at random and is found defective. Find the probability that the item was produced by machine C.

Let $X = \{\text{defective items}\}$. We seek $P(C \mid X)$, the probability that an item is produced by machine C given that the item is defective. By Bayes' theorem,

$$P(C \mid X) = \frac{P(C)\,P(X \mid C)}{P(A)\,P(X \mid A) + P(B)\,P(X \mid B) + P(C)\,P(X \mid C)}$$

$$= \frac{(.10)(.04)}{(.60)(.02) + (.30)(.03) + (.10)(.04)} = \frac{4}{25}$$

4.16. In a certain college, 4% of the men and 1% of the women are taller than 6 feet. Furthermore, 60% of the students are women. Now if a student is selected at random and is taller than 6 feet, what is the probability that the student is a woman?

Let $A = \{\text{students taller than 6 feet}\}$. We seek $P(W \mid A)$, the probability that a student is a woman given that the student is taller than 6 feet. By Bayes' theorem,

$$P(W \mid A) = \frac{P(W)\,P(A \mid W)}{P(W)\,P(A \mid W) + P(M)\,P(A \mid M)} = \frac{(.60)(.01)}{(.60)(.01) + (.40)(.04)} = \frac{3}{11}$$

4.17. Let E be an event for which $P(E) > 0$. Show that the conditional probability function $P(* \mid E)$ satisfies the axioms of a probability space; that is,

[P_1] For any event A, $0 \leq P(A \mid E) \leq 1$.

[P_2] For the certain event S, $P(S \mid E) = 1$.

[P_3] If A and B are mutually exclusive, then $P(A \cup B \mid E) = P(A \mid E) + P(B \mid E)$.

[P_4] If $A_1, A_2, \ldots$ is a sequence of mutually exclusive events, then

$$P(A_1 \cup A_2 \cup \cdots \mid E) = P(A_1 \mid E) + P(A_2 \mid E) + \cdots$$

(i) We have $A \cap E \subset E$; hence $P(A \cap E) \leq P(E)$. Thus $P(A \mid E) = \dfrac{P(A \cap E)}{P(E)} \leq 1$ and is also non-negative. That is, $0 \leq P(A \mid E) \leq 1$ and so **[P_1]** holds.

(ii) We have $S \cap E = E$; hence $P(S \mid E) = \dfrac{P(S \cap E)}{P(E)} = \dfrac{P(E)}{P(E)} = 1$. Thus **[P₂]** holds.

(iii) If A and B are mutually exclusive events, then so are $A \cap E$ and $B \cap E$. Furthermore, $(A \cup B) \cap E = (A \cap E) \cup (B \cap E)$. Thus

$$P((A \cup B) \cap E) \;=\; P((A \cap E) \cup (B \cap E)) \;=\; P(A \cap E) + P(B \cap E)$$

and therefore

$$\begin{aligned} P(A \cup B \mid E) &= \frac{P((A \cup B) \cap E)}{P(E)} = \frac{P(A \cap E) + P(B \cap E)}{P(E)} \\ &= \frac{P(A \cap E)}{P(E)} + \frac{P(B \cap E)}{P(E)} = P(A \mid E) + P(B \mid E) \end{aligned}$$

Hence **[P₃]** holds.

(iv) Similarly if $A_1, A_2, \ldots$ are mutually exclusive, then so are $A_1 \cap E, A_2 \cap E, \ldots$. Thus

$$P((A_1 \cup A_2 \cup \cdots) \cap E) \;=\; P((A_1 \cap E) \cup (A_2 \cap E) \cup \cdots) \;=\; P(A_1 \cap E) + P(A_2 \cap E) + \cdots$$

and therefore

$$\begin{aligned} P(A_1 \cup A_2 \cup \cdots \mid E) &= \frac{P((A_1 \cup A_2 \cup \cdots) \cap E)}{P(E)} = \frac{P(A_1 \cap E) + P(A_2 \cap E) + \cdots}{P(E)} \\ &= \frac{P(A_1 \cap E)}{P(E)} + \frac{P(A_2 \cap E)}{P(E)} + \cdots = P(A_1 \mid E) + P(A_2 \mid E) + \cdots \end{aligned}$$

That is, **[P₄]** holds.

FINITE STOCHASTIC PROCESSES

4.18. A box contains three coins; one coin is fair, one coin is two-headed, and one coin is weighted so that the probability of heads appearing is $\frac{1}{3}$. A coin is selected at random and tossed. Find the probability p that heads appears.

Construct the tree diagram as shown in Figure (*a*) below. Note that I refers to the fair coin, II to the two-headed coin, and III to the weighted coin. Now heads appears along three of the paths; hence

$$p \;=\; \frac{1}{3}\cdot\frac{1}{2} + \frac{1}{3}\cdot 1 + \frac{1}{3}\cdot\frac{1}{3} \;=\; \frac{11}{18}$$

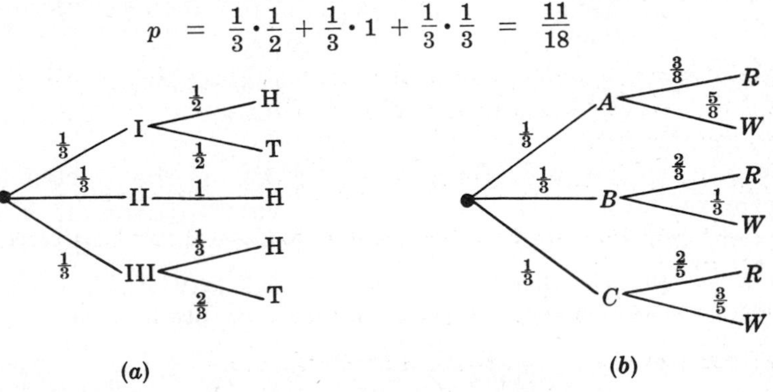

(*a*) (*b*)

4.19. We are given three urns as follows:

Urn A contains 3 red and 5 white marbles.
Urn B contains 2 red and 1 white marble.
Urn C contains 2 red and 3 white marbles.

An urn is selected at random and a marble is drawn from the urn. If the marble is red, what is the probability that it came from urn A?

Construct the tree diagram as shown in Figure (*b*) above.

We seek the probability that A was selected, given that the marble is red; that is, $P(A \mid R)$. In order to find $P(A \mid R)$, it is necessary first to compute $P(A \cap R)$ and $P(R)$.

The probability that urn A is selected and a red marble drawn is $\frac{1}{3}\cdot\frac{3}{8} = \frac{1}{8}$; that is, $P(A \cap R) = \frac{1}{8}$. Since there are three paths leading to a red marble, $P(R) = \frac{1}{3}\cdot\frac{3}{8} + \frac{1}{3}\cdot\frac{2}{3} + \frac{1}{3}\cdot\frac{2}{5} = \frac{173}{360}$. Thus

$$P(A \mid R) = \frac{P(A \cap R)}{P(R)} = \frac{\frac{1}{8}}{\frac{173}{360}} = \frac{45}{173}$$

Alternately, by Bayes' theorem,

$$P(A \mid R) = \frac{P(A)\,P(R \mid A)}{P(A)\,P(R \mid A) + P(B)\,P(R \mid B) + P(C)\,P(R \mid C)}$$

$$= \frac{\frac{1}{3}\cdot\frac{3}{8}}{\frac{1}{3}\cdot\frac{3}{8} + \frac{1}{3}\cdot\frac{2}{3} + \frac{1}{3}\cdot\frac{2}{5}} = \frac{45}{173}$$

4.20. Box A contains nine cards numbered 1 through 9, and box B contains five cards numbered 1 through 5. A box is chosen at random and a card drawn. If the number is even, find the probability that the card came from box A.

The tree diagram of the process is shown in Figure (a) below.

We seek $P(A \mid E)$, the probability that A was selected, given that the number is even. The probability that box A and an even number is drawn is $\frac{1}{2}\cdot\frac{4}{9} = \frac{2}{9}$; that is, $P(A \cap E) = \frac{2}{9}$. Since there are two paths which lead to an even number, $P(E) = \frac{1}{2}\cdot\frac{4}{9} + \frac{1}{2}\cdot\frac{2}{5} = \frac{19}{45}$. Thus

$$P(A \mid E) = \frac{P(A \cap E)}{P(E)} = \frac{\frac{2}{9}}{\frac{19}{45}} = \frac{10}{19}$$

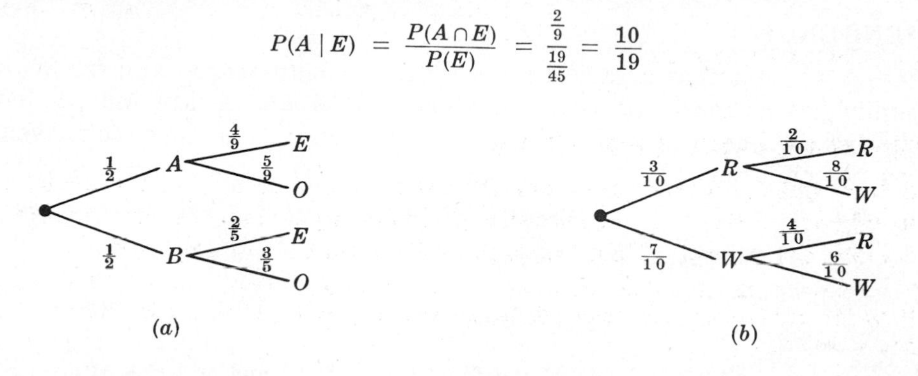

4.21. An urn contains 3 red marbles and 7 white marbles. A marble is drawn from the urn and a marble of the other color is then put into the urn. A second marble is drawn from the urn.

(i) Find the probability p that the second marble is red.

(ii) If both marbles were of the same color, what is the probability p that they were both white?

Construct the tree diagram as shown in Figure (b) above.

(i) Two paths of the tree lead to a red marble: $p = \frac{3}{10}\cdot\frac{2}{10} + \frac{7}{10}\cdot\frac{4}{10} = \frac{17}{50}$.

(ii) The probability that both marbles were white is $\frac{7}{10}\cdot\frac{6}{10} = \frac{21}{50}$. The probability that both marbles were of the same color, i.e. the probability of the reduced sample space, is $\frac{3}{10}\cdot\frac{2}{10} + \frac{7}{10}\cdot\frac{6}{10} = \frac{12}{25}$. Hence the conditional probability $p = \frac{21}{50}/\frac{12}{25} = \frac{7}{8}$.

4.22. We are given two urns as follows:

Urn A contains 3 red and 2 white marbles.
Urn B contains 2 red and 5 white marbles.

An urn is selected at random; a marble is drawn and put into the other urn; then a marble is drawn from the second urn. Find the probability p that both marbles drawn are of the same color.

Construct the following tree diagram:

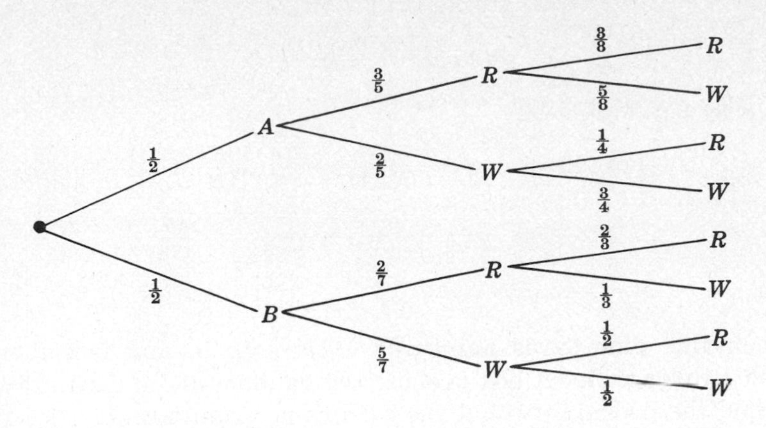

Note that if urn A is selected and a red marble drawn and put into urn B, then urn B has 3 red marbles and 5 white marbles.

Since there are four paths which lead to two marbles of the same color,

$$p = \frac{1}{2}\cdot\frac{3}{5}\cdot\frac{3}{8} + \frac{1}{2}\cdot\frac{2}{5}\cdot\frac{3}{4} + \frac{1}{2}\cdot\frac{2}{7}\cdot\frac{2}{3} + \frac{1}{2}\cdot\frac{5}{7}\cdot\frac{1}{2} = \frac{901}{1680}$$

INDEPENDENCE

4.23. Let A = event that a family has children of both sexes, and let B = event that a family has at most one boy. (i) Show that A and B are independent events if a family has three children. (ii) Show that A and B are dependent events if a family has two children.

(i) We have the equiprobable space $S = \{bbb, bbg, bgb, bgg, gbb, gbg, ggb, ggg\}$. Here

$$A = \{bbg, bgb, bgg, gbb, gbg, ggb\} \quad \text{and so} \quad P(A) = \frac{6}{8} = \frac{3}{4}$$

$$B = \{bgg, gbg, ggb, ggg\} \quad \text{and so} \quad P(B) = \frac{4}{8} = \frac{1}{2}$$

$$A \cap B = \{bgg, gbg, ggb\} \quad \text{and so} \quad P(A \cap B) = \frac{3}{8}$$

Since $P(A)\,P(B) = \frac{3}{4}\cdot\frac{1}{2} = \frac{3}{8} = P(A\cap B)$, A and B are independent.

(ii) We have the equiprobable space $S = \{bb, bg, gb, gg\}$. Here

$$A = \{bg, gb\} \quad \text{and so} \quad P(A) = \frac{1}{2}$$

$$B = \{bg, gb, gg\} \quad \text{and so} \quad P(B) = \frac{3}{4}$$

$$A \cap B = \{bg, gb\} \quad \text{and so} \quad P(A \cap B) = \frac{1}{2}$$

Since $P(A)\,P(B) \neq P(A\cap B)$, A and B are dependent.

4.24. Prove: If A and B are independent events, then A^c and B^c are independent events.

$$\begin{aligned} P(A^c \cap B^c) &= P((A\cup B)^c) = 1 - P(A\cup B) = 1 - P(A) - P(B) + P(A\cap B) \\ &= 1 - P(A) - P(B) + P(A)\,P(B) = [1-P(A)][1-P(B)] = P(A^c)\,P(B^c) \end{aligned}$$

4.25. The probability that a man will live 10 more years is $\frac{1}{4}$, and the probability that his wife will live 10 more years is $\frac{1}{3}$. Find the probability that (i) both will be alive in 10 years, (ii) at least one will be alive in 10 years, (iii) neither will be alive in 10 years, (iv) only the wife will be alive in 10 years.

Let A = event that the man is alive in 10 years, and B = event that his wife is alive in 10 years; then $P(A) = \frac{1}{4}$ and $P(B) = \frac{1}{3}$.

(i) We seek $P(A\cap B)$. Since A and B are independent, $P(A\cap B) = P(A)\,P(B) = \frac{1}{4}\cdot\frac{1}{3} = \frac{1}{12}$.

(ii) We seek $P(A \cup B)$. $P(A \cup B) = P(A) + P(B) - P(A \cap B) = \frac{1}{4} + \frac{1}{3} - \frac{1}{12} = \frac{1}{2}$.

(iii) We seek $P(A^c \cap B^c)$. Now $P(A^c) = 1 - P(A) = 1 - \frac{1}{4} = \frac{3}{4}$ and $P(B^c) = 1 - P(B) = 1 - \frac{1}{3} = \frac{2}{3}$. Furthermore, since A^c and B^c are independent, $P(A^c \cap B^c) = P(A^c)\,P(B^c) = \frac{3}{4} \cdot \frac{2}{3} = \frac{1}{2}$.

Alternately, since $(A \cup B)^c = A^c \cap B^c$, $P(A^c \cap B^c) = P((A \cup B)^c) = 1 - P(A \cup B) = 1 - \frac{1}{2} = \frac{1}{2}$.

(iv) We seek $P(A^c \cap B)$. Since $P(A^c) = 1 - P(A) = \frac{3}{4}$ and A^c and B are independent (see Problem 4.56), $P(A^c \cap B) = P(A^c)\,P(B) = \frac{1}{4}$.

4.26. Box A contains 8 items of which 3 are defective, and box B contains 5 items of which 2 are defective. An item is drawn at random from each box.

(i) What is the probability p that both items are nondefective?

(ii) What is the probability p that one item is defective and one not?

(iii) If one item is defective and one is not, what is the probability p that the defective item came from box A?

(i) The probability of choosing a nondefective item from A is $\frac{5}{8}$ and from B is $\frac{3}{5}$. Since the events are independent, $p = \frac{5}{8} \cdot \frac{3}{5} = \frac{3}{8}$.

(ii) **Method 1.** The probability of choosing two defective items is $\frac{3}{8} \cdot \frac{2}{5} = \frac{3}{20}$. From (i) the probability that both are nondefective is $\frac{3}{8}$. Hence $p = 1 - \frac{3}{8} - \frac{3}{20} = \frac{19}{40}$.

Method 2. The probability p_1 of choosing a defective item from A and a nondefective item from B is $\frac{3}{8} \cdot \frac{3}{5} = \frac{9}{40}$. The probability p_2 of choosing a nondefective item from A and a defective item from B is $\frac{5}{8} \cdot \frac{2}{5} = \frac{1}{4}$. Hence $p = p_1 + p_2 = \frac{9}{40} + \frac{1}{4} = \frac{19}{40}$.

(iii) Consider the events $X = \{\text{defective item from } A\}$ and $Y = \{\text{one item is defective and one nondefective}\}$. We seek $P(X \mid Y)$. By (ii), $P(X \cap Y) = p_1 = \frac{9}{40}$ and $P(Y) = \frac{19}{40}$. Hence

$$p = P(X \mid Y) = \frac{P(X \cap Y)}{P(Y)} = \frac{\frac{9}{40}}{\frac{19}{40}} = \frac{9}{19}$$

4.27. The probabilities that three men hit a target are respectively $\frac{1}{6}$, $\frac{1}{4}$ and $\frac{1}{3}$. Each shoots once at the target. (i) Find the probability p that exactly one of them hits the target. (ii) If only one hit the target, what is the probability that it was the first man?

Consider the events $A = \{\text{first man hits the target}\}$, $B = \{\text{second man hits the target}\}$, and $C = \{\text{third man hits the target}\}$; then $P(A) = \frac{1}{6}$, $P(B) = \frac{1}{4}$ and $P(C) = \frac{1}{3}$. The three events are independent, and $P(A^c) = \frac{5}{6}$, $P(B^c) = \frac{3}{4}$, $P(C^c) = \frac{2}{3}$.

(i) Let $E = \{\text{exactly one man hits the target}\}$. Then

$$E = (A \cap B^c \cap C^c) \cup (A^c \cap B \cap C^c) \cup (A^c \cap B^c \cap C)$$

In other words, if only one hit the target, then it was either only the first man, $A \cap B^c \cap C^c$, or only the second man, $A^c \cap B \cap C^c$, or only the third man, $A^c \cap B^c \cap C$. Since the three events are mutually exclusive, we obtain (using Problem 4.62)

$$\begin{aligned} p = P(E) &= P(A \cap B^c \cap C^c) + P(A^c \cap B \cap C^c) + P(A^c \cap B^c \cap C) \\ &= P(A)\,P(B^c)\,P(C^c) + P(A^c)\,P(B)\,P(C^c) + P(A^c)\,P(B^c)\,P(C) \\ &= \frac{1}{6} \cdot \frac{3}{4} \cdot \frac{2}{3} + \frac{5}{6} \cdot \frac{1}{4} \cdot \frac{2}{3} + \frac{5}{6} \cdot \frac{3}{4} \cdot \frac{1}{3} = \frac{1}{12} + \frac{5}{36} + \frac{5}{24} = \frac{31}{72} \end{aligned}$$

(ii) We seek $P(A \mid E)$, the probability that the first man hit the target given that only one man hit the target. Now $A \cap E = A \cap B^c \cap C^c$ is the event that only the first man hit the target. By (i), $P(A \cap E) = P(A \cap B^c \cap C^c) = \frac{1}{12}$ and $P(E) = \frac{31}{72}$; hence

$$P(A \mid E) = \frac{P(A \cap E)}{P(E)} = \frac{\frac{1}{12}}{\frac{31}{72}} = \frac{6}{31}$$

INDEPENDENT TRIALS

4.28. A certain type of missile hits its target with probability .3. How many missiles should be fired so that there is at least an 80% probability of hitting a target?

The probability of a missile missing its target is .7; hence the probability that n missiles miss a target is $(.7)^n$. Thus we seek the smallest n for which

$$1 - (.7)^n > .8 \quad \text{or equivalently} \quad (.7)^n < .2$$

Compute: $(.7)^1 = .7$, $(.7)^2 = .49$, $(.7)^3 = .343$, $(.7)^4 = .2401$, $(.7)^5 = .16807$. Thus at least 5 missiles should be fired.

4.29. A certain soccer team wins (W) with probability .6, loses (L) with probability .3 and ties (T) with probability .1. The team plays three games over the weekend. (i) Determine the elements of the event A that the team wins at least twice and doesn't lose; and find $P(A)$. (ii) Determine the elements of the event B that the team wins, loses and ties; and find $P(B)$.

(i) A consists of all ordered triples with at least 2 W's and no L's. Thus

$$A = \{\text{WWW, WWT, WTW, TWW}\}$$

Furthermore,

$$\begin{aligned} P(A) &= P(\text{WWW}) + P(\text{WWT}) + P(\text{WTW}) + P(\text{TWW}) \\ &= (.6)(.6)(.6) + (.6)(.6)(.1) + (.6)(.1)(.6) + (.1)(.6)(.6) \\ &= .216 + .036 + .036 + .036 = .324 \end{aligned}$$

(ii) Here $B = \{\text{WLT, WTL, LWT, LTW, TWL, TLW}\}$. Since each element of B has probability $(.6)(.3)(.1) = .018$, $P(B) = 6(.018) = .108$.

4.30. Let S be a finite probability space and let T be the probability space of n independent trials in S. Show that T is well defined; that is, show (i) the probability of each element of T is nonnegative and (ii) the sum of their probabilities is 1.

If $S = \{a_1, \ldots, a_r\}$, then T can be represented by

$$T = \{a_{i_1} \cdots a_{i_n} : i_1, \ldots, i_n = 1, \ldots, r\}$$

Since $P(a_i) \geqq 0$, we have

$$P(a_{i_1} \cdots a_{i_n}) = P(a_{i_1}) \cdots P(a_{i_n}) \geqq 0$$

for a typical element $a_{i_1} \cdots a_{i_n}$ in T, which proves (i)

We prove (ii) by induction on n. It is obviously true for $n = 1$. Therefore we consider $n > 1$ and assume (ii) has been proved for $n - 1$. Then

$$\begin{aligned} \sum_{i_1, \ldots, i_n = 1}^{r} P(a_{i_1} \cdots a_{i_n}) &= \sum_{i_1, \ldots, i_n = 1}^{r} P(a_{i_1}) \cdots P(a_{i_n}) = \sum_{i_1, \ldots, i_{n-1} = 1}^{r} P(a_{i_1}) \cdots P(a_{i_{n-1}}) \sum_{i_n = 1}^{r} P(a_{i_n}) \\ &= \sum_{i_1, \ldots, i_{n-1} = 1}^{r} P(a_{i_1}) \cdots P(a_{i_{n-1}}) = \sum_{i_1, \ldots, i_{n-1} = 1}^{r} P(a_{i_1} \cdots a_{i_{n-1}}) = 1 \end{aligned}$$

by the inductive hypothesis, which proves (ii) for n.

Supplementary Problems

CONDITIONAL PROBABILITY

4.31. A die is tossed. If the number is odd, what is the probability that it is prime?

4.32. Three fair coins are tossed. If both heads and tails appear, determine the probability that exactly one head appears.

4.33. A pair of dice is tossed. If the numbers appearing are different, find the probability that the sum is even.

4.34. A man is dealt 5 red cards from an ordinary deck of 52 cards. What is the probability that they are all of the same suit, i.e. hearts or diamonds?

4.35. A man is dealt 3 spade cards from an ordinary deck of 52 cards. If he is given four more cards, determine the probability that at least two of the additional cards are also spades.

4.36. Two different digits are selected at random from the digits 1 through 9.
(i) If the sum is odd, what is the probability that 2 is one of the numbers selected?
(ii) If 2 is one of the digits selected, what is the probability that the sum is odd?

4.37. Four persons, called North, South, East and West, are each dealt 13 cards from an ordinary deck of 52 cards.
(i) If South has exactly one ace, what is the probability that his partner North has the other three aces?
(ii) If North and South together have 10 hearts, what is the probability that either East or West has the other 3 hearts?

4.38. A class has 10 boys and 5 girls. Three students are selected from the class at random, one after the other. Find the probability that (i) the first two are boys and the third is a girl, (ii) the first and third are boys and the second is a girl, (iii) the first and third are of the same sex, and the second is of the opposite sex.

4.39. In the preceding problem, if the first and third students selected are of the same sex and the second student is of the opposite sex, what is the probability that the second student is a girl?

4.40. In a certain town, 40% of the people have brown hair, 25% have brown eyes, and 15% have both brown hair and brown eyes. A person is selected at random from the town.
(i) If he has brown hair, what is the probability that he also has brown eyes?
(ii) If he has brown eyes, what is the probability that he does not have brown hair?
(iii) What is the probability that he has neither brown hair nor brown eyes?

4.41. Let A and B be events with $P(A) = \frac{1}{3}$, $P(B) = \frac{1}{4}$ and $P(A \cup B) = \frac{1}{2}$. Find (i) $P(A \mid B)$, (ii) $P(B \mid A)$, (iii) $P(A \cap B^c)$, (iv) $P(A \mid B^c)$.

4.42. Let $S = \{a, b, c, d, e, f\}$ with $P(a) = \frac{1}{16}$, $P(b) = \frac{1}{16}$, $P(c) = \frac{1}{8}$, $P(d) = \frac{3}{16}$, $P(e) = \frac{1}{4}$ and $P(f) = \frac{5}{16}$. Let $A = \{a, c, e\}$, $B = \{c, d, e, f\}$ and $C = \{b, c, f\}$. Find (i) $P(A \mid B)$, (ii) $P(B \mid C)$, (iii) $P(C \mid A^c)$, (iv) $P(A^c \mid C)$.

4.43. In a certain college, 25% of the boys and 10% of the girls are studying mathematics. The girls constitute 60% of the student body. If a student is selected at random and is studying mathematics, determine the probability that the student is a girl.

FINITE STOCHASTIC PROCESSES

4.44. We are given two urns as follows:

Urn A contains 5 red marbles, 3 white marbles and 8 blue marbles.

Urn B contains 3 red marbles and 5 white marbles.

A fair die is tossed; if 3 or 6 appears, a marble is chosen from B, otherwise a marble is chosen from A. Find the probability that (i) a red marble is chosen, (ii) a white marble is chosen, (iii) a blue marble is chosen.

4.45. Refer to the preceding problem. (i) If a red marble is chosen, what is the probability that it came from urn A? (ii) If a white marble is chosen, what is the probability that a 5 appeared on the die?

4.46. An urn contains 5 red marbles and 3 white marbles. A marble is selected at random from the urn, discarded, and two marbles of the other color are put into the urn. A second marble is then selected from the urn. Find the probability that (i) the second marble is red, (ii) both marbles are of the same color.

4.47. Refer to the preceding problem. (i) If the second marble is red, what is the probability that the first marble is red? (ii) If both marbles are of the same color, what is the probability that they are both white?

4.48. A box contains three coins, two of them fair and one two-headed. A coin is selected at random and tossed twice. If heads appears both times, what is the probability that the coin is two-headed?

4.49. We are given two urns as follows:

Urn A contains 5 red marbles and 3 white marbles.

Urn B contains 1 red marble and 2 white marbles.

A fair die is tossed; if a 3 or 6 appears, a marble is drawn from B and put into A and then a marble is drawn from A; otherwise, a marble is drawn from A and put into B and then a marble is drawn from B.

(i) What is the probability that both marbles are red?

(ii) What is the probability that both marbles are white?

4.50. Box A contains nine cards numbered 1 through 9, and box B contains five cards numbered 1 through 5. A box is chosen at random and a card drawn; if the card shows an even number, another card is drawn from the same box; if the card shows an odd number, a card is drawn from the other box.

(i) What is the probability that both cards show even numbers?

(ii) If both cards show even numbers, what is the probability that they come from box A?

(iii) What is the probability that both cards show odd numbers?

4.51. A box contains a fair coin and a two-headed coin. A coin is selected at random and tossed. If heads appears, the other coin is tossed; if tails appears, the same coin is tossed.

(i) Find the probability that heads appears on the second toss.

(ii) If heads appeared on the second toss, find the probability that it also appeared on the first toss.

4.52. A box contains three coins, two of them fair and one two-headed. A coin is selected at random and tossed. If heads appears the coin is tossed again; if tails appears, then another coin is selected from the two remaining coins and tossed.

(i) Find the probability that heads appears twice.

(ii) If the same coin is tossed twice, find the probability that it is the two-headed coin.

(iii) Find the probability that tails appears twice.

4.53. Urn A contains x red marbles and y white marbles, and urn B contains z red marbles and v white marbles.

(i) If an urn is selected at random and a marble drawn, what is the probability that the marble is red?

(ii) If a marble is drawn from urn A and put into urn B and then a marble is drawn from urn B, what is the probability that the second marble is red?

4.54. A box contains 5 radio tubes of which 2 are defective. The tubes are tested one after the other until the 2 defective tubes are discovered. What is the probability that the process stopped on the (i) second test, (ii) third test?

4.55. Refer to the preceding problem. If the process stopped on the third test, what is the probability that the first tube is nondefective?

INDEPENDENCE

4.56. Prove: If A and B are independent, then A and B^c are independent and A^c and B are independent.

4.57. Let A and B be events with $P(A) = \frac{1}{4}$, $P(A \cup B) = \frac{1}{3}$ and $P(B) = p$. (i) Find p if A and B are mutually exclusive. (ii) Find p if A and B are independent. (iii) Find p if A is a subset of B.

4.58. Urn A contains 5 red marbles and 3 white marbles, and urn B contains 2 red marbles and 6 white marbles.

(i) If a marble is drawn from each urn, what is the probability that they are both of the same color?

(ii) If two marbles are drawn from each urn, what is the probability that all four marbles are of the same color?

4.59. Let three fair coins be tossed. Let $A = \{\text{all heads or all tails}\}$, $B = \{\text{at least two heads}\}$ and $C = \{\text{at most two heads}\}$. Of the pairs (A, B), (A, C) and (B, C), which are independent and which are dependent?

4.60. The probability that A hits a target is $\frac{1}{4}$ and the probability that B hits a target is $\frac{1}{3}$.

(i) If each fires twice, what is the probability that the target will be hit at least once?

(ii) If each fires once and the target is hit only once, what is the probability that A hit the target?

(iii) If A can fire only twice, how many times must B fire so that there is at least a 90% probability that the target will be hit?

4.61. Let A and B be independent events with $P(A) = \frac{1}{2}$ and $P(A \cup B) = \frac{2}{3}$. Find (i) $P(B)$, (ii) $P(A \mid B)$, (iii) $P(B^c \mid A)$.

4.62. Suppose A, B, C are independent events. Show that any of the combinations

$$A^c, B, C;\ A, B^c, C;\ \ldots;\ A^c, B^c, C;\ \ldots;\ A^c, B^c, C^c$$

are also independent. Furthermore, show that A and $B \cup C$ are independent; and so forth.

INDEPENDENT TRIALS

4.63. A rifleman hits (H) his target with probability .4, and hence misses (M) with probability .6. He fires four times. (i) Determine the elements of the event A that the man hits the target exactly twice; and find $P(A)$. (ii) Find the probability that the man hits the target at least once.

4.64. A team wins (W) with probability .5, loses (L) with probability .3 and ties (T) with probability .2. The team plays twice. (i) Determine the sample space S and the probabilities of the elementary events. (ii) Find the probability that the team wins at least once.

4.65. Consider a countably infinite probability space $S = \{a_1, a_2, \ldots\}$. Let

$$T = S^n = \{(s_1, s_2, \ldots, s_n) : s_i \in S\}$$

and let

$$P(s_1, s_2, \ldots, s_n) = P(s_1)\,P(s_2) \cdots P(s_n)$$

Show that T is also a countably infinite probability space. (This generalizes the definition (page 58) of independent trials to a countably infinite space.)

Answers to Supplementary Problems

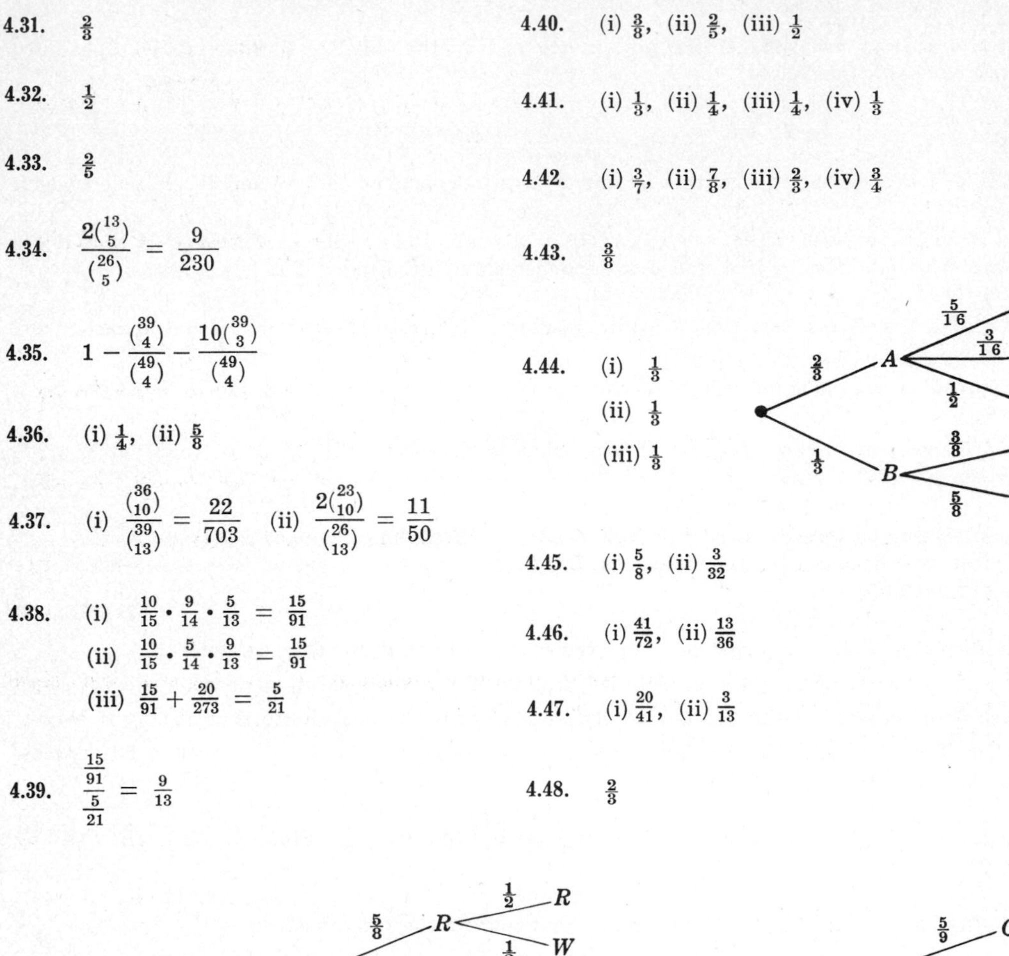

4.31. $\frac{2}{3}$

4.32. $\frac{1}{2}$

4.33. $\frac{2}{5}$

4.34. $\dfrac{2\binom{13}{5}}{\binom{26}{5}} = \dfrac{9}{230}$

4.35. $1 - \dfrac{\binom{39}{4}}{\binom{49}{4}} - \dfrac{10\binom{39}{3}}{\binom{49}{4}}$

4.36. (i) $\frac{1}{4}$, (ii) $\frac{5}{8}$

4.37. (i) $\dfrac{\binom{36}{10}}{\binom{39}{13}} = \dfrac{22}{703}$ (ii) $\dfrac{2\binom{23}{10}}{\binom{26}{13}} = \dfrac{11}{50}$

4.38. (i) $\frac{10}{15} \cdot \frac{9}{14} \cdot \frac{5}{13} = \frac{15}{91}$
(ii) $\frac{10}{15} \cdot \frac{5}{14} \cdot \frac{9}{13} = \frac{15}{91}$
(iii) $\frac{15}{91} + \frac{20}{273} = \frac{5}{21}$

4.39. $\dfrac{\frac{15}{91}}{\frac{5}{21}} = \frac{9}{13}$

4.40. (i) $\frac{3}{8}$, (ii) $\frac{2}{5}$, (iii) $\frac{1}{2}$

4.41. (i) $\frac{1}{3}$, (ii) $\frac{1}{4}$, (iii) $\frac{1}{4}$, (iv) $\frac{1}{3}$

4.42. (i) $\frac{3}{7}$, (ii) $\frac{7}{8}$, (iii) $\frac{2}{3}$, (iv) $\frac{3}{4}$

4.43. $\frac{3}{8}$

4.44. (i) $\frac{1}{3}$
(ii) $\frac{1}{3}$
(iii) $\frac{1}{3}$

4.45. (i) $\frac{5}{8}$, (ii) $\frac{3}{32}$

4.46. (i) $\frac{41}{72}$, (ii) $\frac{13}{36}$

4.47. (i) $\frac{20}{41}$, (ii) $\frac{3}{13}$

4.48. $\frac{2}{3}$

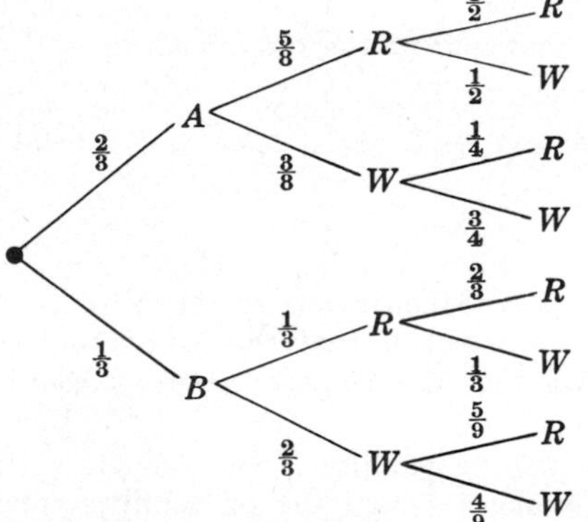

Tree diagram for Problem 4.49

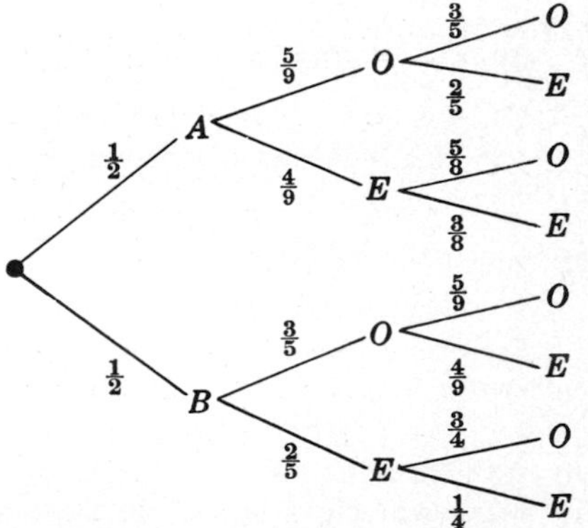

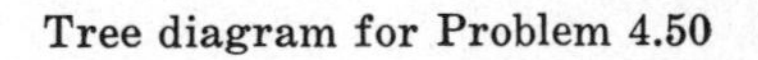

Tree diagram for Problem 4.50

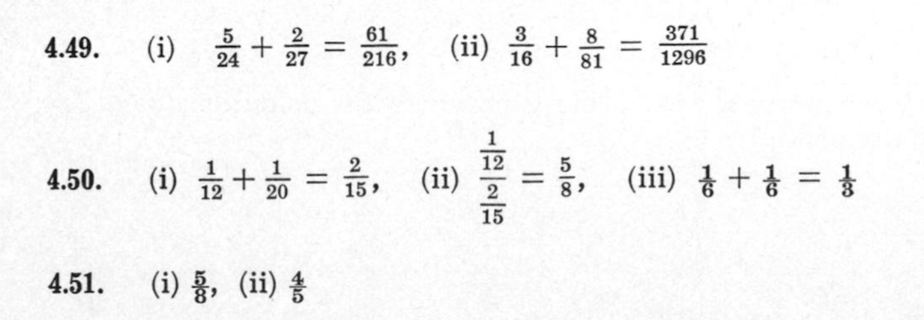

4.49. (i) $\frac{5}{24} + \frac{2}{27} = \frac{61}{216}$, (ii) $\frac{3}{16} + \frac{8}{81} = \frac{371}{1296}$

4.50. (i) $\frac{1}{12} + \frac{1}{20} = \frac{2}{15}$, (ii) $\dfrac{\frac{1}{12}}{\frac{2}{15}} = \frac{5}{8}$, (iii) $\frac{1}{6} + \frac{1}{6} = \frac{1}{3}$

4.51. (i) $\frac{5}{8}$, (ii) $\frac{4}{5}$

4.52. (i) $\frac{1}{12} + \frac{1}{12} + \frac{1}{3} = \frac{1}{2}$, (ii) $\frac{1}{2}$, (iii) $\frac{1}{12}$

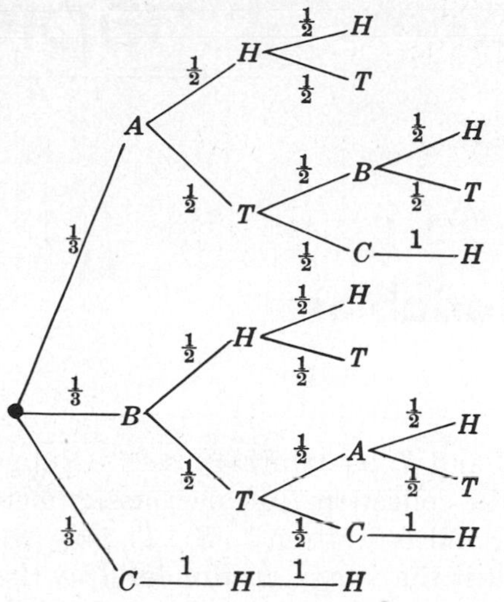

Tree diagram for Problem 4.52

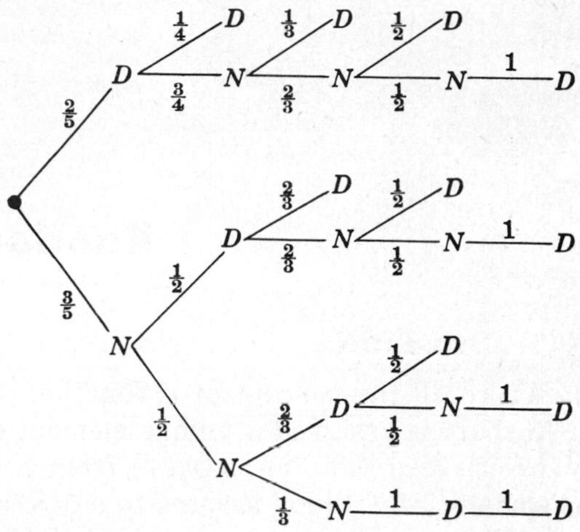

Tree diagram for Problem 4.54

4.53. (i) $\frac{1}{2}(\frac{x}{x+y} + \frac{z}{z+v})$, (ii) $\frac{xz+x+yz}{(x+y)(z+v+1)}$

4.54. (i) $\frac{1}{10}$, (ii) $\frac{3}{10}$; we must include the case where the three nondefective tubes appear first, since the last two tubes must then be the defective ones.

4.55. $\frac{2}{3}$

4.57. (i) $\frac{1}{12}$, (ii) $\frac{1}{9}$, (iii) $\frac{1}{3}$

4.58. (i) $\frac{7}{16}$, (ii) $\frac{55}{784}$

4.59. Only A and B are independent.

4.60. (i) $\frac{3}{4}$, (ii) $\frac{2}{5}$, (iii) 5

4.61. (i) $\frac{1}{3}$, (ii) $\frac{1}{2}$, (iii) $\frac{2}{3}$

4.63. (i) A = {HHMM, HMHM, HMMH, MHHM, MHMH, MMHH}, $P(A) = .3456$
(ii) $1 - (.6)^4 = .8704$

4.64. (i) S = {WW, WL, WT, LW, LL, LT, TW, TL, TT}
(ii) .75

Chapter 5

Random Variables

INTRODUCTION

We recall the concept of a function. Let S and T be arbitrary sets. Suppose to each $s \in S$ there is assigned a unique element of T; the collection f of such assignments is called a *function* (or: *mapping* or *map*) from S into T, and is written $f: S \to T$. We write $f(s)$ for the element of T that f assigns to $s \in S$, and call it the *image* of s under f or the *value* of f at s. The *image* $f(A)$ of any subset A of S, and the *preimage* $f^{-1}(B)$ of any subset B of T are defined by

$$f(A) = \{f(s) : s \in A\} \quad \text{and} \quad f^{-1}(B) = \{s : f(s) \in B\}$$

In words, $f(A)$ consists of the images of points of A and $f^{-1}(B)$ consists of those points whose images belong to B. In particular, the set $f(S)$ of all the image points is called the *image set* (or: *image* or *range*) of f.

Now suppose S is the sample space of some experiment. As noted previously, the outcomes of the experiment, i.e. the sample points of S, need not be numbers. However, we frequently wish to assign a specific number to each outcome, e.g. the sum of the points on a pair of dice, the number of aces in a bridge hand, or the time (in hours) it takes for a lightbulb to burn out. Such an assignment is called a random variable; more precisely,

Definition: A *random variable* X on a sample space S is a function from S into the set $\mathbf{R}$ of real numbers such that the preimage of every interval of $\mathbf{R}$ is an event of S.

We emphasize that if S is a discrete space in which every subset is an event, then every real-valued function on S is a random variable. On the other hand, it can be shown that if S is uncountable then certain real-valued functions on S are not random variables.

If X and Y are random variables on the same sample space S, then $X+Y$, $X+k$, kX and XY (where k is a real number) are the functions on S defined by

$$(X+Y)(s) = X(s) + Y(s) \qquad (kX)(s) = kX(s)$$

$$(X+k)(s) = X(s) + k \qquad (XY)(s) = X(s)\,Y(s)$$

for every $s \in S$. It can be shown that these are also random variables. (This is trivial in the case that every subset of S is an event.)

We use the short notation $P(X=a)$ and $P(a \leq X \leq b)$ for the probability of the events "X maps into a" and "X maps into the interval $[a, b]$." That is,

$$P(X = a) = P(\{s \in S : X(s) = a\})$$

and

$$P(a \leq X \leq b) = P(\{s \in S : a \leq X(s) \leq b\})$$

Analogous meanings are given to $P(X \leq a)$, $P(X=a,\ Y=b)$, $P(a \leq X \leq b,\ c \leq Y \leq d)$, etc.

DISTRIBUTION AND EXPECTATION OF A FINITE RANDOM VARIABLE

Let X be a random variable on a sample space S with a finite image set; say, $X(S) = \{x_1, x_2, \ldots, x_n\}$. We make $X(S)$ into a probability space by defining the probability of x_i to be $P(X = x_i)$ which we write $f(x_i)$. This function f on $X(S)$, i.e. defined by $f(x_i) = P(X = x_i)$, is called the *distribution* or *probability function* of X and is usually given in the form of a table:

x_1	x_2	$\cdots$	x_n
$f(x_1)$	$f(x_2)$	$\cdots$	$f(x_n)$

The distribution f satisfies the conditions

$$\text{(i)}\ f(x_i) \geq 0 \quad \text{and} \quad \text{(ii)}\ \sum_{i=1}^{n} f(x_i) = 1$$

Now if X is a random variable with the above distribution, then the *mean* or *expectation* (or: *expected value*) of X, denoted by $E(X)$ or μ_X, or simply E or μ, is defined by

$$E(X) = x_1 f(x_1) + x_2 f(x_2) + \cdots + x_n f(x_n) = \sum_{i=1}^{n} x_i f(x_i)$$

That is, $E(X)$ is the *weighted average* of the possible values of X, each value weighted by its probability.

Example 5.1: A pair of fair dice is tossed. We obtain the finite equiprobable space S consisting of the 36 ordered pairs of numbers between 1 and 6:

$$S = \{(1,1), (1,2), \ldots, (6,6)\}$$

Let X assign to each point (a, b) in S the maximum of its numbers, i.e. $X(a, b) = \max(a, b)$. Then X is a random variable with image set

$$X(S) = \{1, 2, 3, 4, 5, 6\}$$

We compute the distribution f of X:

$$f(1) = P(X=1) = P(\{(1,1)\}) = \tfrac{1}{36}$$

$$f(2) = P(X=2) = P(\{(2,1), (2,2), (1,2)\}) = \tfrac{3}{36}$$

$$f(3) = P(X=3) = P(\{(3,1), (3,2), (3,3), (2,3), (1,3)\}) = \tfrac{5}{36}$$

$$f(4) = P(X=4) = P(\{(4,1), (4,2), (4,3), (4,4), (3,4), (2,4), (1,4)\}) = \tfrac{7}{36}$$

Similarly,

$$f(5) = P(X=5) = \tfrac{9}{36} \quad \text{and} \quad f(6) = P(X=6) = \tfrac{11}{36}$$

This information is put in the form of a table as follows:

x_i	1	2	3	4	5	6
$f(x_i)$	$\frac{1}{36}$	$\frac{3}{36}$	$\frac{5}{36}$	$\frac{7}{36}$	$\frac{9}{36}$	$\frac{11}{36}$

We next compute the mean of X:

$$E(X) = \sum x_i f(x_i) = 1 \cdot \tfrac{1}{36} + 2 \cdot \tfrac{3}{36} + 3 \cdot \tfrac{5}{36} + 4 \cdot \tfrac{7}{36} + 5 \cdot \tfrac{9}{36} + 6 \cdot \tfrac{11}{36}$$

$$= \tfrac{161}{36} = 4.47$$

Now let Y assign to each point (a, b) in S the sum of its numbers, i.e. $Y(a, b) = a + b$. Then Y is also a random variable on S with image set

$$Y(S) = \{2, 3, 4, 5, 6, 7, 8, 9, 10, 11, 12\}$$

The distribution g of Y follows:

y_i	2	3	4	5	6	7	8	9	10	11	12
$g(y_i)$	$\frac{1}{36}$	$\frac{2}{36}$	$\frac{3}{36}$	$\frac{4}{36}$	$\frac{5}{36}$	$\frac{6}{36}$	$\frac{5}{36}$	$\frac{4}{36}$	$\frac{3}{36}$	$\frac{2}{36}$	$\frac{1}{36}$

We obtain, for example, $g(4) = \frac{3}{36}$ from the fact that $(1,3)$, $(2,2)$, and $(3,1)$ are those points of S for which the sum of the components is 4; hence

$$g(4) = P(Y=4) = P(\{(1,3),(2,2),(3,1)\}) = \tfrac{3}{36}$$

The mean of Y is computed as follows:

$$E(Y) = \sum y_i\, g(y_i) = 2\cdot\tfrac{1}{36} + 3\cdot\tfrac{2}{36} + \cdots + 12\cdot\tfrac{1}{36} = 7$$

The charts which follow graphically describe the above distributions:

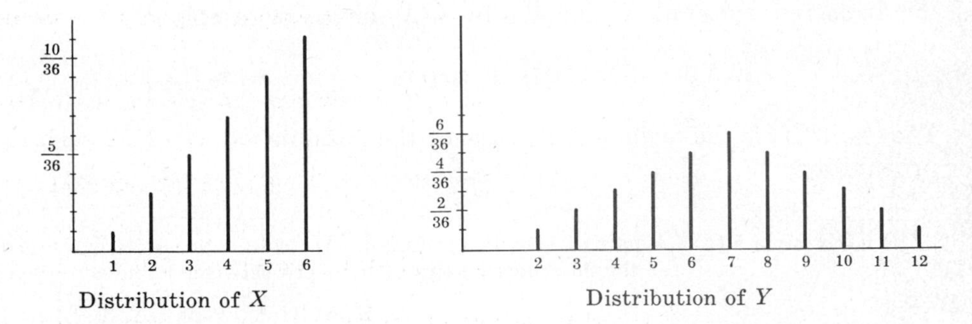

Distribution of X Distribution of Y

Observe that the vertical lines drawn above the numbers on the horizontal axis are proportional to their probabilities.

Example 5.2: A coin weighted so that $P(\text{H}) = \frac{2}{3}$ and $P(\text{T}) = \frac{1}{3}$ is tossed three times. The probabilities of the points in the sample space S = {HHH, HHT, HTH, HTT, THH, THT, TTH, TTT} are as follows:

$$P(\text{HHH}) = \tfrac{2}{3}\cdot\tfrac{2}{3}\cdot\tfrac{2}{3} = \tfrac{8}{27} \qquad P(\text{THH}) = \tfrac{1}{3}\cdot\tfrac{2}{3}\cdot\tfrac{2}{3} = \tfrac{4}{27}$$

$$P(\text{HHT}) = \tfrac{2}{3}\cdot\tfrac{2}{3}\cdot\tfrac{1}{3} = \tfrac{4}{27} \qquad P(\text{THT}) = \tfrac{1}{3}\cdot\tfrac{2}{3}\cdot\tfrac{1}{3} = \tfrac{2}{27}$$

$$P(\text{HTH}) = \tfrac{2}{3}\cdot\tfrac{1}{3}\cdot\tfrac{2}{3} = \tfrac{4}{27} \qquad P(\text{TTH}) = \tfrac{1}{3}\cdot\tfrac{1}{3}\cdot\tfrac{2}{3} = \tfrac{2}{27}$$

$$P(\text{HTT}) = \tfrac{2}{3}\cdot\tfrac{1}{3}\cdot\tfrac{1}{3} = \tfrac{2}{27} \qquad P(\text{TTT}) = \tfrac{1}{3}\cdot\tfrac{1}{3}\cdot\tfrac{1}{3} = \tfrac{1}{27}$$

Let X be the random variable which assigns to each point in S the largest number of successive heads which occurs. Thus,

$$X(\text{TTT}) = 0$$

$$X(\text{HTH}) = 1,\quad X(\text{HTT}) = 1,\quad X(\text{THT}) = 1,\quad X(\text{TTH}) = 1$$

$$X(\text{HHT}) = 2,\quad X(\text{THH}) = 2$$

$$X(\text{HHH}) = 3$$

The image set of X is $X(S) = \{0,1,2,3\}$. We compute the distribution f of X:

$$f(0) = P(\text{TTT}) = \tfrac{1}{27}$$

$$f(1) = P(\{\text{HTH, HTT, THT, TTH}\}) = \tfrac{4}{27} + \tfrac{2}{27} + \tfrac{2}{27} + \tfrac{2}{27} = \tfrac{10}{27}$$

$$f(2) = P(\{\text{HHT, THH}\}) = \tfrac{4}{27} + \tfrac{4}{27} = \tfrac{8}{27}$$

$$f(3) = P(\text{HHH}) = \tfrac{8}{27}$$

This information is put in the form of a table as follows:

x_i	0	1	2	3
$f(x_i)$	$\frac{1}{27}$	$\frac{10}{27}$	$\frac{8}{27}$	$\frac{8}{27}$

The mean of X is computed as follows:

$$E(X) = \sum x_i f(x_i) = 0 \cdot \tfrac{1}{27} + 1 \cdot \tfrac{10}{27} + 2 \cdot \tfrac{8}{27} + 3 \cdot \tfrac{8}{27} = \tfrac{50}{27} = 1.85$$

Example 5.3: A sample of 3 items is selected at random from a box containing 12 items of which 3 are defective. Find the expected number E of defective items.

The sample space S consists of the $\binom{12}{3} = 220$ distinct equally likely samples of size 3. We note that there are:

$$\binom{9}{3} = 84 \text{ samples with no defective items;}$$

$$3 \cdot \binom{9}{2} = 108 \text{ samples with 1 defective item;}$$

$$\binom{3}{2} \cdot 9 = 27 \text{ samples with 2 defective items;}$$

$$\binom{3}{3} = 1 \text{ sample with 3 defective items.}$$

Thus the probability of getting 0, 1, 2 and 3 defective items is respectively 84/220, 108/220, 27/220 and 1/220. Thus the expected number E of defective items is

$$E = 0 \cdot \tfrac{84}{220} + 1 \cdot \tfrac{108}{220} + 2 \cdot \tfrac{27}{220} + 3 \cdot \tfrac{1}{220} = \tfrac{165}{220} = .75$$

Remark: Implicitly we have obtained the expectation of the random variable X which assigns to each sample the number of defective items in the sample.

In a gambling game, the expected value E of the game is considered to be the value of the game to the player. The game is said to be *favorable* to the player if E is positive, and *unfavorable* if E is negative. If $E = 0$, the game is *fair*.

Example 5.4: A player tosses a fair die. If a prime number occurs he wins that number of dollars, but if a non-prime number occurs he loses that number of dollars. The possible outcomes x_i of the game with their respective probabilities $f(x_i)$ are as follows:

x_i	2	3	5	−1	−4	−6
$f(x_i)$	$\frac{1}{6}$	$\frac{1}{6}$	$\frac{1}{6}$	$\frac{1}{6}$	$\frac{1}{6}$	$\frac{1}{6}$

The negative numbers −1, −4 and −6 correspond to the fact that the player loses if a non-prime number occurs. The expected value of the game is

$$E = 2 \cdot \tfrac{1}{6} + 3 \cdot \tfrac{1}{6} + 5 \cdot \tfrac{1}{6} - 1 \cdot \tfrac{1}{6} - 4 \cdot \tfrac{1}{6} - 6 \cdot \tfrac{1}{6} = -\tfrac{1}{6}$$

Thus the game is unfavorable to the player since the expected value is negative.

Our first theorems relate the notion of expectation to operations on random variables.

Theorem 5.1: Let X be a random variable and k a real number. Then (i) $E(kX) = kE(X)$ and (ii) $E(X+k) = E(X) + k$.

Theorem 5.2: Let X and Y be random variables on the same sample space S. Then $E(X+Y) = E(X) + E(Y)$.

A simple induction argument yields

Corollary 5.3: Let $X_1, X_2, \ldots, X_n$ be random variables on S. Then

$$E(X_1 + \cdots + X_n) = E(X_1) + \cdots + E(X_n)$$

VARIANCE AND STANDARD DEVIATION

The mean of a random variable X measures, in a certain sense, the "average" value of X. The next concept, that of the variance of X, measures the "spread" or "dispersion" of X.

Let X be a random variable with the following distribution:

x_1	x_2	$\cdots$	x_n
$f(x_1)$	$f(x_2)$	$\cdots$	$f(x_n)$

Then the *variance* of X, denoted by Var (X), is defined by

$$\text{Var}(X) = \sum_{i=1}^{n} (x_i - \mu)^2 f(x_i) = E((X-\mu)^2)$$

where μ is the mean of X. The *standard deviation* of X, denoted by σ_X, is the (nonnegative) square root of Var (X):

$$\sigma_X = \sqrt{\text{Var}(X)}$$

The next theorem gives us an alternate and sometimes more useful formula for calculating the variance of the random variable X.

Theorem 5.4: $\text{Var}(X) = \sum_{i=1}^{n} x_i^2 f(x_i) - \mu^2 = E(X^2) - \mu^2.$

Proof. Using $\sum x_i f(x_i) = \mu$ and $\sum f(x_i) = 1$, we have

$$\begin{aligned} \sum (x_i - \mu)^2 f(x_i) &= \sum (x_i^2 - 2\mu x_i + \mu^2) f(x_i) \\ &= \sum x_i^2 f(x_i) - 2\mu \sum x_i f(x_i) + \mu^2 \sum f(x_i) \\ &= \sum x_i^2 f(x_i) - 2\mu^2 + \mu^2 = \sum x_i^2 f(x_i) - \mu^2 \end{aligned}$$

which proves the theorem.

Example 5.5: Consider the random variable X of Example 5.1 (which assigns the maximum of the numbers showing on a pair of dice). The distribution of X is

x_i	1	2	3	4	5	6
$f(x_i)$	$\frac{1}{36}$	$\frac{3}{36}$	$\frac{5}{36}$	$\frac{7}{36}$	$\frac{9}{36}$	$\frac{11}{36}$

and its mean is $\mu_X = 4.47$. We compute the variance and standard deviation of X. First we compute $E(X^2)$:

$$\begin{aligned} E(X^2) = \sum x_i^2 f(x_i) &= 1^2 \cdot \tfrac{1}{36} + 2^2 \cdot \tfrac{3}{36} + 3^2 \cdot \tfrac{5}{36} + 4^2 \cdot \tfrac{7}{36} + 5^2 \cdot \tfrac{9}{36} + 6^2 \cdot \tfrac{11}{36} \\ &= \tfrac{791}{36} = 21.97 \end{aligned}$$

Hence

$$\text{Var}(X) = E(X^2) - \mu_X^2 = 21.97 - 19.98 = 1.99 \quad \text{and} \quad \sigma_X = \sqrt{1.99} = 1.4$$

Now consider the random variable Y of Example 5.1 (which assigns the sum of the numbers showing on a pair of dice). The distribution of Y is

y_i	2	3	4	5	6	7	8	9	10	11	12
$g(y_i)$	$\frac{1}{36}$	$\frac{2}{36}$	$\frac{3}{36}$	$\frac{4}{36}$	$\frac{5}{36}$	$\frac{6}{36}$	$\frac{5}{36}$	$\frac{4}{36}$	$\frac{3}{36}$	$\frac{2}{36}$	$\frac{1}{36}$

and its mean is $\mu_Y = 7$. We compute the variance and standard deviation of Y. First we compute $E(Y^2)$:

$$E(Y^2) = \sum y_i^2 g(y_i) = 2^2 \cdot \tfrac{1}{36} + 3^2 \cdot \tfrac{2}{36} + \cdots + 12^2 \cdot \tfrac{1}{36} = \tfrac{1974}{36} = 54.8$$

Hence

$$\text{Var}(Y) = E(Y^2) - \mu_Y^2 = 54.8 - 49 = 5.8 \quad \text{and} \quad \sigma_Y = \sqrt{5.8} = 2.4$$

We establish some properties of the variance in

Theorem 5.5: Let X be a random variable and k a real number. Then (i) $\text{Var}(X+k) = \text{Var}(X)$ and (ii) $\text{Var}(kX) = k^2\text{Var}(X)$. Hence $\sigma_{X+k} = \sigma_X$ and $\sigma_{kX} = |k|\sigma_X$.

Remark 1. There is a physical interpretation of mean and variance. Suppose at each point x_i on the x axis there is placed a unit with mass $f(x_i)$. Then the mean is the center of gravity of the system, and the variance is the moment of inertia of the system.

Remark 2. Many random variables give rise to the same distribution; hence we frequently speak of the mean, variance and standard deviation of a distribution instead of the underlying random variable.

Remark 3. Let X be a random variable with mean μ and standard deviation $\sigma > 0$. The *standardized random variable* X^* corresponding to X is defined by

$$X^* = \frac{X - \mu}{\sigma}$$

We show (Problem 5.23) that $E(X^*) = 0$ and $\text{Var}(X^*) = 1$.

JOINT DISTRIBUTION

Let X and Y be random variables on a sample space S with respective image sets

$$X(S) = \{x_1, x_2, \ldots, x_n\} \quad \text{and} \quad Y(S) = \{y_1, y_2, \ldots, y_m\}$$

We make the product set

$$X(S) \times Y(S) = \{(x_1, y_1), (x_1, y_2), \ldots, (x_n, y_m)\}$$

into a probability space by defining the *probability* of the ordered pair (x_i, y_j) to be $P(X = x_i, Y = y_j)$ which we write $h(x_i, y_j)$. This function h on $X(S) \times Y(S)$, i.e. defined by $h(x_i, y_j) = P(X = x_i, Y = y_j)$, is called the *joint distribution* or *joint probability function* of X and Y and is usually given in the form of a table:

X \ Y	y_1	y_2	$\cdots$	y_m	Sum
x_1	$h(x_1, y_1)$	$h(x_1, y_2)$	$\cdots$	$h(x_1, y_m)$	$f(x_1)$
x_2	$h(x_2, y_1)$	$h(x_2, y_2)$	$\cdots$	$h(x_2, y_m)$	$f(x_2)$
$\cdots$	$\cdots$	$\cdots$	$\cdots$	$\cdots$	$\cdots$
x_n	$h(x_n, y_1)$	$h(x_n, y_2)$	$\cdots$	$h(x_n, y_m)$	$f(x_n)$
Sum	$g(y_1)$	$g(y_2)$	$\cdots$	$g(y_m)$	

The above functions f and g are defined by

$$f(x_i) = \sum_{j=1}^{m} h(x_i, y_j) \quad \text{and} \quad g(y_j) = \sum_{i=1}^{n} h(x_i, y_j)$$

i.e. $f(x_i)$ is the sum of the entries in the ith row and $g(y_j)$ is the sum of the entries in the jth column; they are called the *marginal distributions* and are, in fact, the (individual) distributions of X and Y respectively (Problem 5.12). The joint distribution h satisfies the conditions

$$\text{(i)}\ \ h(x_i, y_j) \geqq 0 \quad \text{and} \quad \text{(ii)}\ \ \sum_{i=1}^{n} \sum_{j=1}^{m} h(x_i, y_j) = 1$$

Now if X and Y are random variables with the above joint distribution (and respective means μ_X and μ_Y), then the *covariance* of X and Y, denoted by Cov (X, Y), is defined by

$$\text{Cov}(X, Y) = \sum_{i,j} (x_i - \mu_X)(y_j - \mu_Y)\, h(x_i, y_j) = E[(X - \mu_X)(Y - \mu_Y)]$$

or equivalently (see Problem 5.18) by

$$\text{Cov}(X, Y) = \sum_{i,j} x_i y_j\, h(x_i, y_j) - \mu_X \mu_Y = E(XY) - \mu_X \mu_Y$$

The *correlation* of X and Y, denoted by $\rho(X, Y)$, is defined by

$$\rho(X, Y) = \frac{\text{Cov}(X, Y)}{\sigma_X \sigma_Y}$$

The correlation ρ is dimensionless and has the following properties:

(i) $\rho(X, Y) = \rho(Y, X)$ (iii) $\rho(X, X) = 1,\ \rho(X, -X) = -1$

(ii) $-1 \leqq \rho \leqq 1$ (iv) $\rho(aX + b, cY + d) = \rho(X, Y)$, if $a, c \neq 0$

We show below (Example 5.7) that pairs of random variables with identical (individual) distributions can have distinct covariances and correlations. Thus Cov (X, Y) and $\rho(X, Y)$ are measurements of the way that X and Y are interrelated.

Example 5.6: A pair of fair dice is tossed. We obtain the finite equiprobable space S consisting of the 36 ordered pairs of numbers between 1 and 6:

$$S = \{(1,1), (1,2), \ldots, (6,6)\}$$

Let X and Y be the random variables on S in Example 5.1, i.e. X assigns the maximum of the numbers and Y the sum of the numbers to each point of S. The joint distribution of X and Y follows:

$X \backslash Y$	2	3	4	5	6	7	8	9	10	11	12	Sum
1	$\frac{1}{36}$	0	0	0	0	0	0	0	0	0	0	$\frac{1}{36}$
2	0	$\frac{2}{36}$	$\frac{1}{36}$	0	0	0	0	0	0	0	0	$\frac{3}{36}$
3	0	0	$\frac{2}{36}$	$\frac{2}{36}$	$\frac{1}{36}$	0	0	0	0	0	0	$\frac{5}{36}$
4	0	0	0	$\frac{2}{36}$	$\frac{2}{36}$	$\frac{2}{36}$	$\frac{1}{36}$	0	0	0	0	$\frac{7}{36}$
5	0	0	0	0	$\frac{2}{36}$	$\frac{2}{36}$	$\frac{2}{36}$	$\frac{2}{36}$	$\frac{1}{36}$	0	0	$\frac{9}{36}$
6	0	0	0	0	0	$\frac{2}{36}$	$\frac{2}{36}$	$\frac{2}{36}$	$\frac{2}{36}$	$\frac{2}{36}$	$\frac{1}{36}$	$\frac{11}{36}$
Sum	$\frac{1}{36}$	$\frac{2}{36}$	$\frac{3}{36}$	$\frac{4}{36}$	$\frac{5}{36}$	$\frac{6}{36}$	$\frac{5}{36}$	$\frac{4}{36}$	$\frac{3}{36}$	$\frac{2}{36}$	$\frac{1}{36}$	

The above entry $h(3,5) = \frac{2}{36}$ comes from the fact that $(3,2)$ and $(2,3)$ are the only points in S whose maximum number is 3 and whose sum is 5; hence

$$h(3,5) = P(X=3, Y=5) = P(\{(3,2),(2,3)\}) = \tfrac{2}{36}$$

The other entries are obtained in a similar manner.

We compute the covariance and correlation of X and Y. First we compute $E(XY)$:

$$\begin{aligned} E(XY) &= \sum x_i y_j\, h(x_i, y_j) \\ &= 1\cdot 2\cdot \tfrac{1}{36} + 2\cdot 3\cdot \tfrac{2}{36} + 2\cdot 4\cdot \tfrac{1}{36} + \cdots + 6\cdot 12\cdot \tfrac{1}{36} \\ &= \tfrac{1232}{36} = 34.2 \end{aligned}$$

By Example 5.1, $\mu_X = 4.47$ and $\mu_Y = 7$, and by Example 5.5, $\sigma_X = 1.4$ and $\sigma_Y = 2.4$; hence

$$\operatorname{Cov}(X,Y) = E(XY) - \mu_X\mu_Y = 34.2 - (4.47)(7) = 2.9$$

and
$$\rho(X,Y) = \frac{\operatorname{Cov}(X,Y)}{\sigma_X\sigma_Y} = \frac{2.9}{(1.4)(2.4)} = .86$$

Example 5.7: Let X and Y, and X' and Y' be random variables with the following joint distributions:

X \ Y	4	10	Sum
1	$\frac{1}{4}$	$\frac{1}{4}$	$\frac{1}{2}$
3	$\frac{1}{4}$	$\frac{1}{4}$	$\frac{1}{2}$
Sum	$\frac{1}{2}$	$\frac{1}{2}$	

X' \ Y'	4	10	Sum
1	0	$\frac{1}{2}$	$\frac{1}{2}$
3	$\frac{1}{2}$	0	$\frac{1}{2}$
Sum	$\frac{1}{2}$	$\frac{1}{2}$	

Observe that X and X', and Y and Y' have identical distributions:

x_i	1	3
$f(x_i)$	$\frac{1}{2}$	$\frac{1}{2}$

Distribution of X and X'

y_i	4	10
$g(y_i)$	$\frac{1}{2}$	$\frac{1}{2}$

Distribution of Y and Y'

We show that $\operatorname{Cov}(X,Y) \neq \operatorname{Cov}(X',Y')$ and hence $\rho(X,Y) \neq \rho(X',Y')$. We first compute $E(XY)$ and $E(X'Y')$:

$$\begin{aligned} E(XY) &= 1\cdot 4\cdot \tfrac{1}{4} + 1\cdot 10\cdot \tfrac{1}{4} + 3\cdot 4\cdot \tfrac{1}{4} + 3\cdot 10\cdot \tfrac{1}{4} = 14 \\ E(X'Y') &= 1\cdot 4\cdot 0 + 1\cdot 10\cdot \tfrac{1}{2} + 3\cdot 4\cdot \tfrac{1}{2} + 3\cdot 10\cdot 0 = 11 \end{aligned}$$

Since $\mu_X = \mu_{X'} = 2$ and $\mu_Y = \mu_{Y'} = 7$,

$$\operatorname{Cov}(X,Y) = E(XY) - \mu_X\mu_Y = 0 \quad \text{and} \quad \operatorname{Cov}(X',Y') = E(X'Y') - \mu_{X'}\mu_{Y'} = -3$$

Remark: The notion of a joint distribution h is extended to any finite number of random variables $X, Y, \ldots, Z$ in the obvious way; that is, h is a function on the product set $X(S) \times Y(S) \times \cdots \times Z(S)$ defined by

$$h(x_i, y_j, \ldots, z_k) = P(X = x_i,\ Y = y_j,\ \ldots,\ Z = z_k)$$

INDEPENDENT RANDOM VARIABLES

A finite number of random variables $X, Y, \ldots, Z$ on a sample space S are said to be *independent* if

$$P(X = x_i, Y = y_j, \ldots, Z = z_k) = P(X = x_i)\,P(Y = y_j)\cdots P(Z = z_k)$$

for any values $x_i, y_j, \ldots, z_k$. In particular, X and Y are independent if

$$P(X = x_i,\ Y = y_j) \quad = \quad P(X = x_i)\, P(Y = y_j)$$

Now if X and Y have respective distributions f and g, and joint distribution h, then the above equation can be written as

$$h(x_i, y_j) \quad = \quad f(x_i)\, g(y_j)$$

In other words, X and Y are independent if each entry $h(x_i, y_j)$ is the product of its marginal entries.

Example 5.8: Let X and Y be random variables with the following joint distribution:

X \ Y	2	3	4	Sum
1	.06	.15	.09	.30
2	.14	.35	.21	.70
Sum	.20	.50	.30	

Thus the distributions of X and Y are as follows:

x	1	2
$f(x)$	.30	.70

Distribution of X

y	2	3	4
$g(y)$	.20	.50	.30

Distribution of Y

X and Y are independent random variables since each entry of the joint distribution can be obtained by multiplying its marginal entries; that is,

$$P(X = x_i,\ Y = y_j) \quad = \quad P(X = x_i)\, P(Y = y_j)$$

for each i and each j.

We establish some important properties of independent random variables which do not hold in general; namely,

Theorem 5.6: Let X and Y be independent random variables. Then:

(i) $E(XY) \ = \ E(X)E(Y)$,
(ii) $\mathrm{Var}\,(X+Y) \ = \ \mathrm{Var}\,(X) + \mathrm{Var}\,(Y)$,
(iii) $\mathrm{Cov}\,(X, Y) = 0$.

Part (ii) in the above theorem generalizes to the very important

Theorem 5.7: Let $X_1, X_2, \ldots, X_n$ be independent random variables. Then

$$\mathrm{Var}\,(X_1 + \cdots + X_n) \quad = \quad \mathrm{Var}\,(X_1) + \cdots + \mathrm{Var}\,(X_n)$$

FUNCTIONS OF A RANDOM VARIABLE

Let X and Y be random variables on the same sample space S. Then Y is said to be a *function* of X if Y can be represented $Y = \Phi(X)$ for some real-valued function Φ of a real variable; that is, if $Y(s) = \Phi[X(s)]$ for every $s \in S$. For example, kX, X^2, $X + k$ and $(X+k)^2$ are all functions of X with $\Phi(x) = kx,\ x^2,\ x + k$ and $(x+k)^2$ respectively. We have the fundamental

Theorem 5.8: Let X and Y be random variables on the same sample space S with $Y = \Phi(X)$. Then

$$E(Y) = \sum_{i=1}^{n} \Phi(x_i)\, f(x_i)$$

where f is the distribution function of X.

Similarly, a random variable Z is said to be a function of X and Y if Z can be represented $Z = \Phi(X, Y)$ where Φ is a real-valued function of two real variables; that is, if

$$Z(s) = \Phi[X(s), Y(s)]$$

for every $s \in S$. Corresponding to the above theorem, we have

Theorem 5.9: Let X, Y and Z be random variables on the same sample space S with $Z = \Phi(X, Y)$. Then

$$E(Z) = \sum_{i,j} \Phi(x_i, y_j)\, h(x_i, y_j)$$

where h is the joint distribution of X and Y.

We remark that the above two theorems have been used implicitly in the preceding discussion and theorems. We also remark that the proof of Theorem 5.9 is given as a supplementary problem, and that the theorem generalizes to a function of n random variables in the obvious way.

DISCRETE RANDOM VARIABLES IN GENERAL

Now suppose X is a random variable on S with a countably infinite image set; say $X(S) = \{x_1, x_2, \ldots\}$. Such random variables together with those with finite image sets (considered above) are called *discrete* random variables. As in the finite case, we make $X(S)$ into a probability space by defining the *probability* of x_i to be $f(x_i) = P(X = x_i)$ and call f the *distribution* of X:

x_1	x_2	x_3	$\cdots$
$f(x_1)$	$f(x_2)$	$f(x_3)$	$\cdots$

The *expectation* $E(X)$ and *variance* $\mathrm{Var}(X)$ are defined by

$$E(X) = x_1 f(x_1) + x_2 f(x_2) + \cdots = \sum_{i=1}^{\infty} x_i f(x_i)$$

$$\mathrm{Var}(X) = (x_1 - \mu)^2 f(x_1) + (x_2 - \mu)^2 f(x_2) + \cdots = \sum_{i=1}^{\infty} (x_i - \mu)^2 f(x_i)$$

when the relevant series converge absolutely. It can be shown that $\mathrm{Var}(X)$ exists if and only if $\mu = E(X)$ and $E(X^2)$ both exist and that in this case the formula

$$\mathrm{Var}(X) = E(X^2) - \mu^2$$

is valid just as in the finite case. When $\mathrm{Var}(X)$ exists, the *standard deviation* σ_X is defined as in the finite case by

$$\sigma_X = \sqrt{\mathrm{Var}(X)}$$

The notions of joint distribution, independent random variables and functions of random variables carry over directly to the general case. It can be shown that if X and Y are defined on the same sample space S and if $\mathrm{Var}(X)$ and $\mathrm{Var}(Y)$ both exist, then the series

$$\operatorname{Cov}(X, Y) = \sum_{i,j} (x_i - \mu_X)(y_j - \mu_Y)\, h(x_i, y_j)$$

converges absolutely and the relation

$$\operatorname{Cov}(X, Y) = \sum_{i,j} x_i y_j\, h(x_i, y_j) - \mu_X \mu_Y = E(XY) - \mu_X \mu_Y$$

holds just as in the finite case.

Remark: To avoid technicalities we will establish many theorems in this chapter only for finite random variables.

CONTINUOUS RANDOM VARIABLES

Suppose that X is a random variable whose image set $X(S)$ is a continuum of numbers such as an interval. Recall from the definition of random variables that the set $\{a \leq X \leq b\}$ is an event in S and therefore the probability $P(a \leq X \leq b)$ is well defined. We assume that there is a piecewise continuous function $f: \mathbf{R} \to \mathbf{R}$ such that $P(a \leq X \leq b)$ is equal to the area under the graph of f between $x = a$ and $x = b$ (as shown on the right). In the language of calculus,

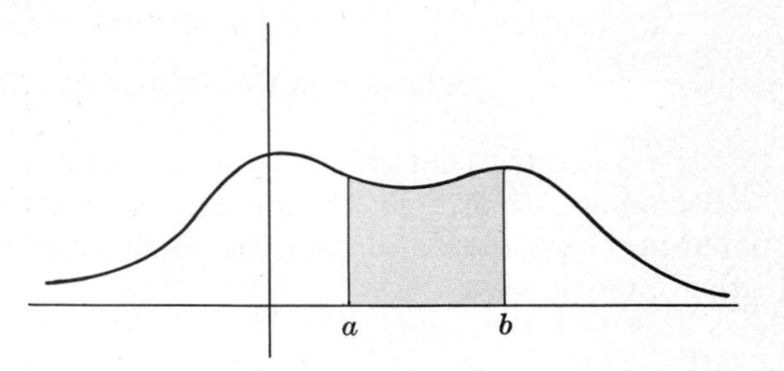

$P(a \leq X \leq b)$ = area of shaded region

$$P(a \leq X \leq b) = \int_a^b f(x)\, dx$$

In this case X is said to be a *continuous random variable.* The function f is called the *distribution* or the *continuous probability function* (or: *density function*) of X; it satisfies the conditions

$$\text{(i)}\quad f(x) \geq 0 \quad \text{and} \quad \text{(ii)}\quad \int_{\mathbf{R}} f(x)\, dx = 1$$

That is, f is nonnegative and the total area under its graph is 1.

The *expectation* $E(X)$ is defined by

$$E(X) = \int_{\mathbf{R}} x\, f(x)\, dx$$

when it exists. Functions of random variables are defined just as in the discrete case; and it can be shown that if $Y = \Phi(X)$, then

$$E(Y) = \int_{\mathbf{R}} \Phi(x)\, f(x)\, dx$$

when the right side exists. The *variance* $\operatorname{Var}(X)$ is defined by

$$\operatorname{Var}(X) = E((X-\mu)^2) = \int_{\mathbf{R}} (x-\mu)^2 f(x)\, dx$$

when it exists. Just as in the discrete case, it can be shown that $\operatorname{Var}(X)$ exists if and only if $\mu = E(X)$ and $E(X^2)$ both exist and then

$$\operatorname{Var}(X) = E(X^2) - \mu^2 = \int_{\mathbf{R}} x^2 f(x)\, dx - \mu^2$$

The *standard deviation* σ_X is defined by $\sigma_X = \sqrt{\text{Var}(X)}$ when Var (X) exists.

We have already remarked that we will establish many results for finite random variables and take them for granted in the general discrete case and in the continuous case.

Example 5.9: Let X be a continuous random variable with the following distribution:

$$f(x) = \begin{cases} \frac{1}{2}x & \text{if } 0 \leq x \leq 2 \\ 0 & \text{elsewhere} \end{cases}$$

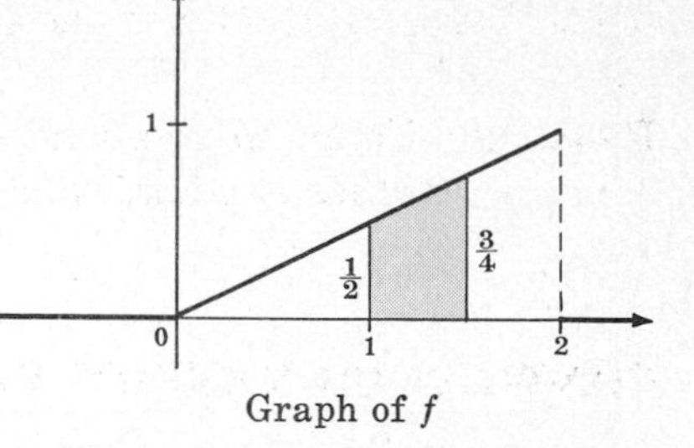

Graph of f

Then

$$P(1 \leq X \leq 1.5) = \text{area of shaded region in diagram} = \tfrac{1}{2} \cdot \tfrac{1}{2}(\tfrac{1}{2} + \tfrac{3}{4}) = \tfrac{5}{16}$$

We next compute the expectation, variance and standard deviation of X:

$$E(X) = \int_{\mathbf{R}} x\, f(x)\, dx = \int_0^2 \tfrac{1}{2}x^2\, dx = \left[\frac{x^3}{6}\right]_0^2 = \frac{4}{3}$$

$$E(X^2) = \int_{\mathbf{R}} x^2 f(x)\, dx = \int_0^2 \tfrac{1}{2}x^3\, dx = \left[\frac{x^4}{8}\right]_0^2 = 2$$

$$\text{Var}(X) = E(X^2) - \mu^2 = 2 - \frac{16}{9} = \frac{2}{9} \quad \text{and} \quad \sigma_X = \sqrt{\frac{2}{9}} = \frac{1}{3}\sqrt{2}$$

A finite number of continuous random variables, say $X, Y, \ldots, Z$, are said to be *independent* if for any intervals $[a, a'], [b, b'], \ldots, [c, c']$,

$$P(a \leq X \leq a', b \leq Y \leq b', \ldots, c \leq Z \leq c') = P(a \leq X \leq a')P(b \leq Y \leq b') \cdots P(c \leq Z \leq c')$$

Observe that intervals play the same role in the continuous case as points did in the discrete case.

CUMULATIVE DISTRIBUTION FUNCTION

Let X be a random variable (discrete or continuous). The *cumulative distribution function* F of X is the function $F: \mathbf{R} \to \mathbf{R}$ defined by

$$F(a) = P(X \leq a)$$

If X is a discrete random variable with distribution f, then F is the "step function" defined by

$$F(x) = \sum_{x_i \leq x} f(x_i)$$

On the other hand, if X is a continuous random variable with distribution f, then

$$F(x) = \int_{-\infty}^{x} f(t)\, dt$$

In either case, F is monotonic increasing, i.e.

$$F(a) \leq F(b) \quad \text{whenever} \quad a \leq b$$

and the limit of F to the left is 0 and to the right is 1:

$$\lim_{x \to -\infty} F(x) = 0 \quad \text{and} \quad \lim_{x \to \infty} F(x) = 1$$

Example 5.10: Let X be a discrete random variable with the following distribution:

x_i	-2	1	2	4
$f(x_i)$	$\frac{1}{4}$	$\frac{1}{8}$	$\frac{1}{2}$	$\frac{1}{8}$

The graph of the cumulative distribution function F of X follows:

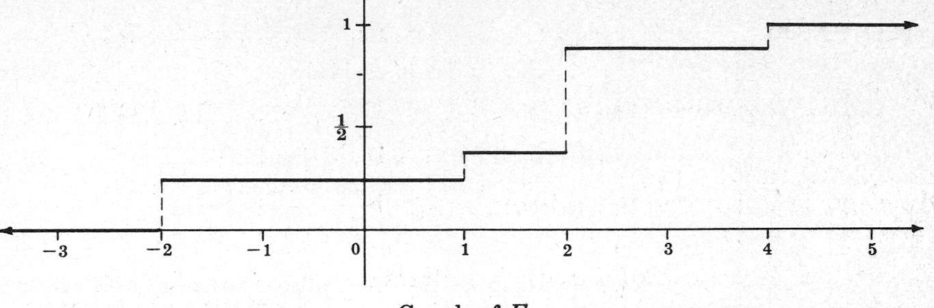

Graph of F

Observe that F is a "step function" with a step at the x_i with height $f(x_i)$.

Example 5.11: Let X be a continuous random variable with the following distribution:

$$f(x) = \begin{cases} \frac{1}{2}x & \text{if } 0 \leq x \leq 2 \\ 0 & \text{elsewhere} \end{cases}$$

The cumulative distribution function F and its graph follows:

$$F(x) = \begin{cases} 0 & \text{if } x < 0 \\ \frac{1}{4}x^2 & \text{if } 0 \leq x \leq 2 \\ 1 & \text{if } x > 2 \end{cases}$$

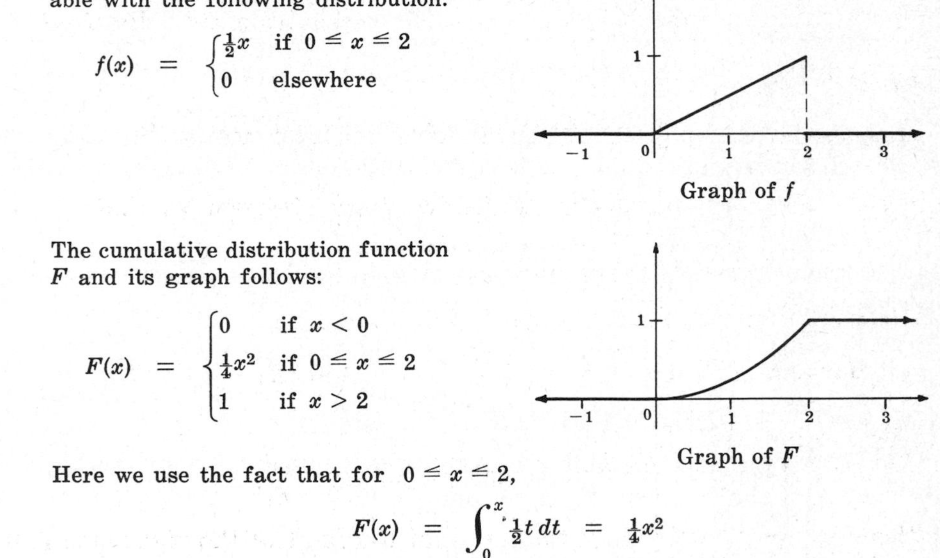

Graph of f

Graph of F

Here we use the fact that for $0 \leq x \leq 2$,

$$F(x) = \int_0^x \tfrac{1}{2}t\,dt = \tfrac{1}{4}x^2$$

TCHEBYCHEFF'S INEQUALITY. LAW OF LARGE NUMBERS

The intuitive idea of probability is the so-called "law of averages", i.e. if an event A occurs with probability p then the "average number of occurrences of A" approaches p as the number of (independent) trials increases. This concept is made precise by the Law of Large Numbers stated below. The proof of this theorem uses the well-known Tchebycheff's inequality which follows:

Theorem 5.10 (Tchebycheff's inequality): Let X be a random variable with mean μ and standard deviation σ. Then for every $\epsilon > 0$,

$$P(|X - \mu| \geq \epsilon) \leq \frac{\sigma^2}{\epsilon^2}$$

Proof. We begin with the definition of variance:

$$\sigma^2 = \text{Var}(X) = \sum_i (x_i - \mu)^2 f(x_i)$$

We delete all the terms in the above series for which $|x_i - \mu| < \epsilon$. This does not increase the value of the series, since all its terms are nonnegative; that is,

$$\sigma^2 \geqq \sum_i{}^* (x_i - \mu)^2 f(x_i)$$

where the asterisk indicates that the summation extends only over those i for which $|x_i - \mu| \geqq \epsilon$. Thus this new summation does not increase in value if we replace each $|x_i - \mu|$ by ϵ; that is,

$$\sigma^2 \geqq \sum_i{}^* \epsilon^2 f(x_i) = \epsilon^2 \sum_i{}^* f(x_i)$$

But $\sum^* f(x_i)$ is equal to the probability that $|X - \mu| \geqq \epsilon$; hence

$$\sigma^2 \geqq \epsilon^2 P(|X - \mu| \geqq \epsilon)$$

Dividing by ϵ^2 we get the desired inequality.

Theorem 5.11 (Law of Large Numbers): Let $X_1, X_2, \ldots$ be a sequence of independent random variables with the same distribution with mean μ and variance σ^2. Let

$$\bar{S}_n = (X_1 + X_2 + \cdots + X_n)/n$$

(called the *sample mean*). Then for any $\epsilon > 0$

$$\lim_{n \to \infty} P(|\bar{S}_n - \mu| \geqq \epsilon) = 0 \quad \text{or equivalently} \quad \lim_{n \to \infty} P(|\bar{S}_n - \mu| < \epsilon) = 1$$

Proof. Note first that

$$E(\bar{S}_n) = \frac{E(X_1) + E(X_2) + \cdots + E(X_n)}{n} = \frac{n\mu}{n} = \mu$$

Since $X_1, \ldots, X_n$ are independent, it follows from Theorem 5.7 that

$$\operatorname{Var}(X_1 + \cdots + X_n) = \operatorname{Var}(X_1) + \cdots + \operatorname{Var}(X_n) = n\sigma^2$$

Therefore by Theorem 5.5(ii),

$$\operatorname{Var}(\bar{S}_n) = \operatorname{Var}\left(\frac{X_1 + \cdots + X_n}{n}\right) = \frac{1}{n^2}\operatorname{Var}(X_1 + \cdots + X_n) = \frac{1}{n^2}(n\sigma^2) = \frac{\sigma^2}{n}$$

Thus by Tchebycheff's inequality,

$$P(|\bar{S}_n - \mu| \geqq \epsilon) \leqq \frac{\sigma^2}{n\epsilon^2}$$

The theorem now follows from the fact that the limit of the right side is 0 as $n \to \infty$.

The following remarks are in order.

Remark 1. We proved Tchebycheff's inequality only for the discrete case. The continuous case follows from an analogous proof which uses integrals instead of summations.

Remark 2. We proved the Law of Large Numbers only in the case that the variance of the X_i exists, i.e. does not diverge. We note that the theorem is true whenever $E(X_i)$ exists.

Remark 3. The above Law of Large Numbers is also called the Weak Law of Large Numbers because of a similar, but stronger, theorem called the Strong Law of Large Numbers.

Solved Problems

RANDOM VARIABLES AND EXPECTATION

5.1. Find the expectation μ, variance σ^2 and standard deviation σ of each of the following distributions:

(i)

x_i	2	3	11
$f(x_i)$	$\frac{1}{3}$	$\frac{1}{2}$	$\frac{1}{6}$

(ii)

x_i	−5	−4	1	2
$f(x_i)$	$\frac{1}{4}$	$\frac{1}{8}$	$\frac{1}{2}$	$\frac{1}{8}$

(iii)

x_i	1	3	4	5
$f(x_i)$	.4	.1	.2	.3

(i)
$$\mu = \sum x_i f(x_i) = 2\cdot\tfrac{1}{3} + 3\cdot\tfrac{1}{2} + 11\cdot\tfrac{1}{6} = 4$$
$$\sum x_i^2 f(x_i) = 2^2\cdot\tfrac{1}{3} + 3^2\cdot\tfrac{1}{2} + 11^2\cdot\tfrac{1}{6} = 26$$
$$\sigma^2 = \sum x_i^2 f(x_i) - \mu^2 = 26 - 16 = 10$$
$$\sigma = \sqrt{10} = 3.2$$

(ii)
$$\mu = \sum x_i f(x_i) = -5\cdot\tfrac{1}{4} - 4\cdot\tfrac{1}{8} + 1\cdot\tfrac{1}{2} + 2\cdot\tfrac{1}{8} = -1$$
$$\sum x_i^2 f(x_i) = 25\cdot\tfrac{1}{4} + 16\cdot\tfrac{1}{8} + 1\cdot\tfrac{1}{2} + 4\cdot\tfrac{1}{8} = 9.25$$
$$\sigma^2 = \sum x_i^2 f(x_i) - \mu^2 = 9.25 - 1 = 8.25$$
$$\sigma = \sqrt{8.25} = 2.9$$

(iii)
$$\mu = \sum x_i f(x_i) = 1(.4) + 3(.1) + 4(.2) + 5(.3) = 3$$
$$\sum x_i^2 f(x_i) = 1(.4) + 9(.1) + 16(.2) + 25(.3) = 12$$
$$\sigma^2 = \sum x_i^2 f(x_i) - \mu^2 = 12 - 9 = 3$$
$$\sigma = \sqrt{3} = 1.7$$

5.2. A fair die is tossed. Let X denote twice the number appearing, and let Y denote 1 or 3 according as an odd or an even number appears. Find the distribution, expectation, variance and standard deviation of (i) X, (ii) Y, (iii) $X+Y$, (iv) XY.

The sample space is $S = \{1,2,3,4,5,6\}$, and each number appears with probability $\frac{1}{6}$.

(i) $X(1) = 2$, $X(2) = 4$, $X(3) = 6$, $X(4) = 8$, $X(5) = 10$, $X(6) = 12$. Thus $X(S) = \{2,4,6,8,10,12\}$ and each number has probability $\frac{1}{6}$. Thus the distribution of X is as follows:

x_i	2	4	6	8	10	12
$f(x_i)$	$\frac{1}{6}$	$\frac{1}{6}$	$\frac{1}{6}$	$\frac{1}{6}$	$\frac{1}{6}$	$\frac{1}{6}$

Accordingly,

$$\mu_X = E(X) = \sum x_i f(x_i)$$
$$= 2\cdot\tfrac{1}{6} + 4\cdot\tfrac{1}{6} + 6\cdot\tfrac{1}{6} + 8\cdot\tfrac{1}{6} + 10\cdot\tfrac{1}{6} + 12\cdot\tfrac{1}{6} = \tfrac{42}{6} = 7$$

$$E(X^2) = \sum x_i^2 f(x_i)$$
$$= 4\cdot\tfrac{1}{6} + 16\cdot\tfrac{1}{6} + 36\cdot\tfrac{1}{6} + 64\cdot\tfrac{1}{6} + 100\cdot\tfrac{1}{6} + 144\cdot\tfrac{1}{6} = \tfrac{364}{6} = 60.7$$

$$\sigma_X^2 = \text{Var}(X) = E(X^2) - \mu_X^2 = 60.7 - (7)^2 = 11.7$$

$$\sigma_X = \sqrt{11.7} = 3.4$$

(ii) $Y(1) = 1,\ Y(2) = 3,\ Y(3) = 1,\ Y(4) = 3,\ Y(5) = 1,\ Y(6) = 3$. Hence $Y(S) = \{1, 3\}$ and

$$g(1) = P(Y=1) = P(\{1,3,5\}) = \tfrac{3}{6} = \tfrac{1}{2} \quad \text{and} \quad g(3) = P(Y=3) = P(\{2,4,6\}) = \tfrac{3}{6} = \tfrac{1}{2}$$

Thus the distribution of Y is as follows:

y_j	1	3
$g(y_j)$	$\frac{1}{2}$	$\frac{1}{2}$

Accordingly,

$$\mu_Y = E(Y) = \sum y_j g(y_j) = 1 \cdot \tfrac{1}{2} + 3 \cdot \tfrac{1}{2} = 2$$

$$E(Y^2) = \sum y_j^2 g(y_j) = 1 \cdot \tfrac{1}{2} + 9 \cdot \tfrac{1}{2} = 5$$

$$\sigma_Y^2 = \operatorname{Var}(Y) = E(Y^2) - \mu_Y^2 = 5 - (2)^2 = 1$$

$$\sigma_Y = \sqrt{1} = 1$$

(iii) Using $(X+Y)(s) = X(s) + Y(s)$, we obtain

$(X+Y)(1) = 2 + 1 = 3$ $\quad$ $(X+Y)(3) = 6 + 1 = 7$ $\quad$ $(X+Y)(5) = 10 + 1 = 11$

$(X+Y)(2) = 4 + 3 = 7$ $\quad$ $(X+Y)(4) = 8 + 3 = 11$ $\quad$ $(X+Y)(6) = 12 + 3 = 15$

Hence the image set is $(X+Y)(S) = \{3, 7, 11, 15\}$ and 3 and 15 occur with probability $\frac{1}{6}$, and 7 and 11 with probability $\frac{2}{6}$. That is, the distribution of $X + Y$ is as follows:

z_i	3	7	11	15
$p(z_i)$	$\frac{1}{6}$	$\frac{2}{6}$	$\frac{2}{6}$	$\frac{1}{6}$

Thus

$$E(X+Y) = 3 \cdot \tfrac{1}{6} + 7 \cdot \tfrac{2}{6} + 11 \cdot \tfrac{2}{6} + 15 \cdot \tfrac{1}{6} = \tfrac{54}{6} = 9$$

$$E((X+Y)^2) = 9 \cdot \tfrac{1}{6} + 49 \cdot \tfrac{2}{6} + 121 \cdot \tfrac{2}{6} + 225 \cdot \tfrac{1}{6} = \tfrac{574}{6} = 95.7$$

$$\operatorname{Var}(X+Y) = E((X+Y)^2) - \mu^2 = 95.7 - 9^2 = 14.7$$

$$\sigma_{X+Y} = \sqrt{14.7} = 3.8$$

Observe that $E(X) + E(Y) = 7 + 2 = 9 = E(X+Y)$, but $\operatorname{Var}(X) + \operatorname{Var}(Y) = 11.7 + 1 = 12.7 \neq \operatorname{Var}(X+Y)$.

(iv) Using $(XY)(s) = X(s)\,Y(s)$, we obtain

$(XY)(1) = 2 \cdot 1 = 2$ $\quad$ $(XY)(3) = 6 \cdot 1 = 6$ $\quad$ $(XY)(5) = 10 \cdot 1 = 10$

$(XY)(2) = 4 \cdot 3 = 12$ $\quad$ $(XY)(4) = 8 \cdot 3 = 24$ $\quad$ $(XY)(6) = 12 \cdot 3 = 36$

Hence the distribution of XY is as follows:

w_i	2	6	10	12	24	36
$p(w_i)$	$\frac{1}{6}$	$\frac{1}{6}$	$\frac{1}{6}$	$\frac{1}{6}$	$\frac{1}{6}$	$\frac{1}{6}$

Thus

$$E(XY) = 2 \cdot \tfrac{1}{6} + 6 \cdot \tfrac{1}{6} + 10 \cdot \tfrac{1}{6} + 12 \cdot \tfrac{1}{6} + 24 \cdot \tfrac{1}{6} + 36 \cdot \tfrac{1}{6} = \tfrac{90}{6} = 15$$

$$E((XY)^2) = 4 \cdot \tfrac{1}{6} + 36 \cdot \tfrac{1}{6} + 100 \cdot \tfrac{1}{6} + 144 \cdot \tfrac{1}{6} + 576 \cdot \tfrac{1}{6} + 1296 \cdot \tfrac{1}{6}$$
$$= \tfrac{2156}{6} = 359.3$$

$$\operatorname{Var}(XY) = E((XY)^2) - \mu^2 = 359.3 - 15^2 = 134.3$$

$$\sigma_{XY} = \sqrt{134.3} = 11.6$$

5.3. A coin weighted so that $P(\mathrm{H}) = \frac{3}{4}$ and $P(\mathrm{T}) = \frac{1}{4}$ is tossed three times. Let X be the random variable which denotes the longest string of heads which occurs. Find the distribution, expectation, variance and standard deviation of X.

The random variable X is defined on the sample space

$$S = \{\mathrm{HHH, HHT, HTH, HTT, THH, THT, TTH, TTT}\}$$

The points in S have the following respective probabilities:

$$P(\mathrm{HHH}) = \tfrac{3}{4}\cdot\tfrac{3}{4}\cdot\tfrac{3}{4} = \tfrac{27}{64} \qquad P(\mathrm{THH}) = \tfrac{1}{4}\cdot\tfrac{3}{4}\cdot\tfrac{3}{4} = \tfrac{9}{64}$$
$$P(\mathrm{HHT}) = \tfrac{3}{4}\cdot\tfrac{3}{4}\cdot\tfrac{1}{4} = \tfrac{9}{64} \qquad P(\mathrm{THT}) = \tfrac{1}{4}\cdot\tfrac{3}{4}\cdot\tfrac{1}{4} = \tfrac{3}{64}$$
$$P(\mathrm{HTH}) = \tfrac{3}{4}\cdot\tfrac{1}{4}\cdot\tfrac{3}{4} = \tfrac{9}{64} \qquad P(\mathrm{TTH}) = \tfrac{1}{4}\cdot\tfrac{1}{4}\cdot\tfrac{3}{4} = \tfrac{3}{64}$$
$$P(\mathrm{HTT}) = \tfrac{3}{4}\cdot\tfrac{1}{4}\cdot\tfrac{1}{4} = \tfrac{3}{64} \qquad P(\mathrm{TTT}) = \tfrac{1}{4}\cdot\tfrac{1}{4}\cdot\tfrac{1}{4} = \tfrac{1}{64}$$

Since X denotes the longest string of heads,

$$X(\mathrm{TTT}) = 0; \quad X(\mathrm{HTT}) = 1,\ X(\mathrm{HTH}) = 1,\ X(\mathrm{THT}) = 1,\ X(\mathrm{TTH}) = 1;$$
$$X(\mathrm{HHT}) = 2,\ X(\mathrm{THH}) = 2; \quad X(\mathrm{HHH}) = 3$$

Thus the image set of X is $X(S) = \{0, 1, 2, 3\}$. The probability $f(x_i)$ of each number x_i in $X(S)$ is obtained by summing the probabilities of the points in S whose image is x_i:

$$f(0) = P(\mathrm{TTT}) = \tfrac{1}{64}$$
$$f(1) = P(\mathrm{HTT}) + P(\mathrm{HTH}) + P(\mathrm{THT}) + P(\mathrm{TTH}) = \tfrac{18}{64}$$
$$f(2) = P(\mathrm{HHT}) + P(\mathrm{THH}) = \tfrac{18}{64}$$
$$f(3) = P(\mathrm{HHH}) = \tfrac{27}{64}$$

Accordingly, the distribution of X is as follows:

x_i	0	1	2	3
$f(x_i)$	$\frac{1}{64}$	$\frac{18}{64}$	$\frac{18}{64}$	$\frac{27}{64}$

Thus

$$\mu = E(X) = 0\cdot\tfrac{1}{64} + 1\cdot\tfrac{18}{64} + 2\cdot\tfrac{18}{64} + 3\cdot\tfrac{27}{64} = \tfrac{135}{64} = 2.1$$
$$E(X^2) = 0\cdot\tfrac{1}{64} + 1\cdot\tfrac{18}{64} + 4\cdot\tfrac{18}{64} + 9\cdot\tfrac{27}{64} = \tfrac{333}{64} = 5.2$$
$$\sigma^2 = \mathrm{Var}(X) = E(X^2) - \mu^2 = 5.2 - (2.1)^2 = .8$$
$$\sigma = \sqrt{.8} = .9$$

5.4. A fair coin is tossed until a head or five tails occurs. Find the expected number E of tosses of the coin.

Only one toss occurs if heads occurs the first time, i.e. the event H. Two tosses occur if the first is tails and the second is heads, i.e. the event TH. Three tosses occur if the first two are tails and the third is heads, i.e. the event TTH. Four tosses occur if TTTH occurs, and five tosses occur if either TTTTH or TTTTT occurs. Hence

$$f(1) = P(\mathrm{H}) = \tfrac{1}{2}$$
$$f(2) = P(\mathrm{TH}) = \tfrac{1}{4}$$
$$f(3) = P(\mathrm{TTH}) = \tfrac{1}{8}$$
$$f(4) = P(\mathrm{TTTH}) = \tfrac{1}{16}$$
$$f(5) = P(\mathrm{TTTTH}) + P(\mathrm{TTTTT}) = \tfrac{1}{32} + \tfrac{1}{32} = \tfrac{1}{16}$$

Accordingly, $E = 1\cdot\tfrac{1}{2} + 2\cdot\tfrac{1}{4} + 3\cdot\tfrac{1}{8} + 4\cdot\tfrac{1}{16} + 5\cdot\tfrac{1}{16} = \tfrac{31}{16} = 1.9.$

5.5. Concentric circles of radius 1 and 3 inches are drawn on a circular target of radius 5 inches. A man receives 10, 5 or 3 points according if he hits the target inside the smaller circle, inside the middle annular region or inside the outer annular region respectively. Suppose the man hits the target with probability $\frac{1}{2}$ and then is just as likely to hit one point of the target as the other. Find the expected number E of points he scores each time he fires.

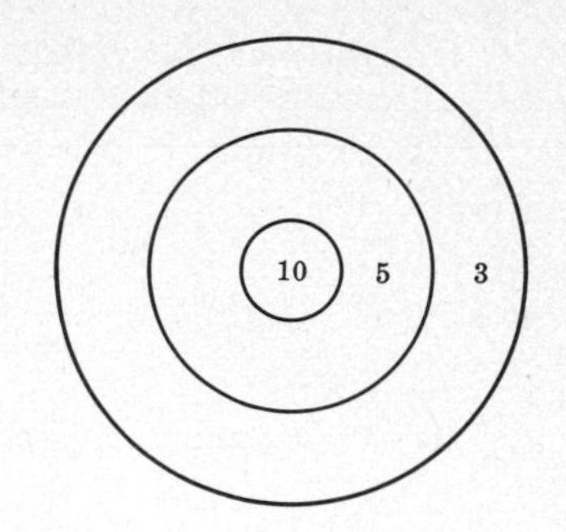

The probability of scoring 10, 5, 3 or 0 points follows:

$$f(10) \;=\; \frac{1}{2}\cdot\frac{\text{area of 10 points}}{\text{area of target}} \;=\; \frac{1}{2}\cdot\frac{\pi(1)^2}{\pi(5)^2} \;=\; \frac{1}{50}$$

$$f(5) \;=\; \frac{1}{2}\cdot\frac{\text{area of 5 points}}{\text{area of target}} \;=\; \frac{1}{2}\cdot\frac{\pi(3)^2-\pi(1)^2}{\pi(5)^2} \;=\; \frac{8}{50}$$

$$f(3) \;=\; \frac{1}{2}\cdot\frac{\text{area of 3 points}}{\text{area of target}} \;=\; \frac{1}{2}\cdot\frac{\pi(5)^2-\pi(3)^2}{\pi(5)^2} \;=\; \frac{16}{50}$$

$$f(0) \;=\; \frac{1}{2}$$

Thus $E = 10\cdot\frac{1}{50} + 5\cdot\frac{8}{50} + 3\cdot\frac{16}{50} + 0\cdot\frac{1}{2} = \frac{98}{50} = 1.96$.

5.6. A player tosses two fair coins. He wins \$1 or \$2 according as 1 or 2 heads appear. On the other hand, he loses \$5 if no heads appear. Determine the expected value E of the game and if it is favorable to the player.

The probability that 2 heads appear is $\frac{1}{4}$, that 2 tails (no heads) appear is $\frac{1}{4}$ and that 1 head appears is $\frac{1}{2}$. Thus the probability of winning \$2 is $\frac{1}{4}$, of winning \$1 is $\frac{1}{2}$, and of losing \$5 is $\frac{1}{4}$. Hence $E = 2\cdot\frac{1}{4} + 1\cdot\frac{1}{2} - 5\cdot\frac{1}{4} = -\frac{1}{4} = -0.25$. That is, the expected value of the game is minus 25¢, and so is unfavorable to the player.

5.7. A player tosses two fair coins. He wins \$5 if 2 heads occur, \$2 if 1 head occurs and \$1 if no heads occur. (i) Find his expected winnings. (ii) How much should he pay to play the game if it is to be fair?

(i) The probability of winning \$5 is $\frac{1}{4}$, of winning \$2 is $\frac{1}{2}$, and of winning \$1 is $\frac{1}{4}$; hence $E = 5\cdot\frac{1}{4} + 2\cdot\frac{1}{2} + 1\cdot\frac{1}{4} = 2.50$, that is, the expected winnings are \$2.50.

(ii) If he pays \$2.50 to play the game, then the game is fair.

JOINT DISTRIBUTIONS, INDEPENDENT RANDOM VARIABLES

5.8. Suppose X and Y have the following joint distribution:

X \ Y	-3	2	4	Sum
1	.1	.2	.2	.5
3	.3	.1	.1	.5
Sum	.4	.3	.3	

(i) Find the distributions of X and Y.
(ii) Find Cov (X, Y), i.e. the covariance of X and Y.
(iii) Find $\rho(X, Y)$, i.e. the correlation of X and Y.
(iv) Are X and Y independent random variables?

(i) The marginal distribution on the right is the distribution of X, and the marginal distribution on the bottom is the distribution of Y. Namely,

x_i	1	3
$f(x_i)$	.5	.5

Distribution of X

y_j	−3	2	4
$g(y_j)$	.4	.3	.3

Distribution of Y

(ii) First compute μ_X and μ_Y:

$$\mu_X = \sum x_i f(x_i) = (1)(.5) + (3)(.5) = 2$$

$$\mu_Y = \sum y_j g(y_j) = (-3)(.4) + (2)(.3) + (4)(.3) = .6$$

Next compute $E(XY)$:

$$E(XY) = \sum x_i y_j h(x_i, y_j)$$

$$= (1)(-3)(.1) + (1)(2)(.2) + (1)(4)(.2) + (3)(-3)(.3) + (3)(2)(.1) + (3)(4)(.1) = 0$$

Then $\text{Cov}(X, Y) = E(XY) - \mu_X\mu_Y = 0 - (2)(.6) = -1.2$

(iii) First compute σ_X and σ_Y:

$$E(X^2) = \sum x_i^2 f(x_i) = (1)(.5) + (9)(.5) = 5$$

$$\sigma_X^2 = \text{Var}(X) = E(X^2) - \mu_X^2 = 5 - (2)^2 = 1$$

$$\sigma_X = \sqrt{1} = 1$$

and

$$E(Y^2) = \sum y_j^2 g(y_j) = (9)(.4) + (4)(.3) + (16)(.3) = 9.6$$

$$\sigma_Y^2 = \text{Var}(Y) = E(Y^2) - \mu_Y^2 = 9.6 - (.6)^2 = 9.24$$

$$\sigma_Y = \sqrt{9.24} = 3.0$$

Then

$$\rho(X, Y) = \frac{\text{Cov}(X, Y)}{\sigma_X \sigma_Y} = \frac{-1.2}{(1)(3.0)} = -.4$$

(iv) X and Y are not independent, since $P(X = 1, Y = -3) \neq P(X = 1)\,P(Y = -3)$, i.e. the entry $h(1, -3) = .1$ is not equal to $f(1)\,g(-3) = (.5)(.4) = .2$, the product of its marginal entries.

5.9. Let X and Y be independent random variables with the following distributions:

x_i	1	2
$f(x_i)$	.6	.4

Distribution of X

y_j	5	10	15
$g(y_j)$	.2	.5	.3

Distribution of Y

Find the joint distribution h of X and Y.

Since X and Y are independent, the joint distribution h can be obtained from the marginal distributions f and g. First construct the joint distribution table with only the marginal distributions as shown below on the left, and then multiply the marginal entries to obtain the other entries, i.e. set $h(x_i, y_j) = f(x_i)\,g(y_j)$, as shown below on the right.

X \ Y	5	10	15	Sum
1				.6
2				.4
Sum	.2	.5	.3	

X \ Y	5	10	15	Sum
1	.12	.30	.18	.6
2	.08	.20	.12	.4
Sum	.2	.5	.3	

5.10. A fair coin is tossed three times. Let X denote 0 or 1 according as a head or a tail occurs on the first toss, and let Y denote the number of heads which occur. Determine (i) the distributions of X and Y, (ii) the joint distribution h of X and Y, (iii) $\text{Cov}(X, Y)$.

(i) The sample space S consists of the following eight points, each with probability $\frac{1}{8}$:

$$S = \{\text{HHH, HHT, HTH, HTT, THH, THT, TTH, TTT}\}$$

We have

$$X(\text{HHH}) = 0,\ X(\text{HHT}) = 0,\ X(\text{HTH}) = 0,\ X(\text{HTT}) = 0$$
$$X(\text{THH}) = 1,\ X(\text{THT}) = 1,\ X(\text{TTH}) = 1,\ X(\text{TTT}) = 1$$

and

$$Y(\text{HHH}) = 3$$
$$Y(\text{HHT}) = 2,\ Y(\text{HTH}) = 2,\ Y(\text{THH}) = 2$$
$$Y(\text{HTT}) = 1,\ Y(\text{THT}) = 1,\ Y(\text{TTH}) = 1$$
$$Y(\text{TTT}) = 0$$

Thus the distributions of X and Y are as follows:

x_i	0	1
$f(x_i)$	$\frac{1}{2}$	$\frac{1}{2}$

Distribution of X

y_j	0	1	2	3
$g(y_j)$	$\frac{1}{8}$	$\frac{3}{8}$	$\frac{3}{8}$	$\frac{1}{8}$

Distribution of Y

(ii) The joint distribution h of X and Y follows:

X \ Y	0	1	2	3	Sum
0	0	$\frac{1}{8}$	$\frac{2}{8}$	$\frac{1}{8}$	$\frac{1}{2}$
1	$\frac{1}{8}$	$\frac{2}{8}$	$\frac{1}{8}$	0	$\frac{1}{2}$
Sum	$\frac{1}{8}$	$\frac{3}{8}$	$\frac{3}{8}$	$\frac{1}{8}$	

We obtain, for example, the entry $h(0, 2) = P(X = 0, Y = 2) = P(\{\text{HTH, HHT}\}) = \frac{2}{8}$.

(iii)

$$\mu_X = \sum x_i f(x_i) = 0 \cdot \tfrac{1}{2} + 1 \cdot \tfrac{1}{2} = \tfrac{1}{2}$$

$$\mu_Y = \sum y_j g(y_j) = 0 \cdot \tfrac{1}{8} + 1 \cdot \tfrac{3}{8} + 2 \cdot \tfrac{3}{8} + 3 \cdot \tfrac{1}{8} = \tfrac{3}{2}$$

$$E(XY) = \sum x_i y_j h(x_i, y_j) = 1 \cdot 1 \cdot \tfrac{2}{8} + 1 \cdot 2 \cdot \tfrac{1}{8} + \text{terms with a factor } 0 = \tfrac{1}{2}$$

$$\text{Cov}(X, Y) = E(XY) - \mu_X\mu_Y = \tfrac{1}{2} - \tfrac{1}{2} \cdot \tfrac{3}{2} = -\tfrac{1}{4}$$

5.11. Let X be a random variable with the following distribution and let $Y = X^2$:

x_i	-2	-1	1	2
$f(x_i)$	$\frac{1}{4}$	$\frac{1}{4}$	$\frac{1}{4}$	$\frac{1}{4}$

Determine (i) the distribution g of Y, (ii) the joint distribution h of X and Y, (iii) $\text{Cov}(X, Y)$ and $\rho(X, Y)$.

(i) Since $Y = X^2$, the random variable Y can only take on the values 4 and 1. Furthermore,

$$g(4) = P(Y=4) = P(X=2 \text{ or } X=-2) = P(X=2) + P(X=-2) = \tfrac{1}{4} + \tfrac{1}{4} = \tfrac{1}{2}$$

and, similarly, $g(1) = \frac{1}{2}$. Hence the distribution g of Y is as follows:

y_j	1	4
$g(y_j)$	$\frac{1}{2}$	$\frac{1}{2}$

(ii) The joint distribution h of X and Y appears below. Note that if $X = -2$, then $Y = 4$; hence $h(-2,1) = 0$ and $h(-2,4) = f(-2) = \frac{1}{4}$. The other entries are obtained in a similar way.

X \ Y	1	4	Sum
−2	0	$\frac{1}{4}$	$\frac{1}{4}$
−1	$\frac{1}{4}$	0	$\frac{1}{4}$
1	$\frac{1}{4}$	0	$\frac{1}{4}$
2	0	$\frac{1}{4}$	$\frac{1}{4}$
Sum	$\frac{1}{2}$	$\frac{1}{2}$	

(iii)

$$\mu_X = E(X) = \sum x_i f(x_i) = -2 \cdot \tfrac{1}{4} - 1 \cdot \tfrac{1}{4} + 1 \cdot \tfrac{1}{4} + 2 \cdot \tfrac{1}{4} = 0$$

$$\mu_Y = E(Y) = \sum y_j g(y_j) = 1 \cdot \tfrac{1}{2} + 4 \cdot \tfrac{1}{2} = \tfrac{5}{2}$$

$$E(XY) = \sum x_i y_j h(x_i, y_j) = -8 \cdot \tfrac{1}{4} - 1 \cdot \tfrac{1}{4} + 1 \cdot \tfrac{1}{4} + 8 \cdot \tfrac{1}{4} = 0$$

$$\text{Cov}(X,Y) = E(XY) - \mu_X \mu_Y = 0 - 0 \cdot \tfrac{5}{2} = 0 \quad \text{and so } \rho(X,Y) = 0$$

Remark: This example shows that although Y is a function of X it is still possible for the covariance and correlation of X and Y to be 0, as in the case when X and Y are independent (Theorem 5.6). Notice, however, that X and Y are not independent in this example.

PROOFS OF THEOREMS

Remark: In all the proofs, X and Y are random variables with distributions f and g respectively and joint distribution h.

5.12. Show that $f(x_i) = \sum_j h(x_i, y_j)$ and $g(y_j) = \sum_i h(x_i, y_j)$, i.e. that the marginal distributions are the (individual) distributions of X and Y.

Let $A_i = \{X = x_i\}$ and $B_j = \{Y = y_j\}$; that is, let $A_i = X^{-1}(x_i)$ and $B_j = Y^{-1}(y_j)$. Thus the B_j are disjoint and $S = \cup_j B_j$. Hence

$$A_i = A_i \cap S = A_i \cap (\cup_j B_j) = \cup_j (A_i \cap B_j)$$

where the $A_i \cap B_j$ are also disjoint. Accordingly,

$$f(x_i) = P(X = x_i) = P(A_i) = \sum_j P(A_i \cap B_j) = \sum_j P(X = x_i, Y = y_j) = \sum_j h(x_i, y_j)$$

The proof for g is similar.

5.13. Prove Theorem 5.8: Let X and Y be random variables on the same sample space S with $Y = \Phi(X)$. Then $E(Y) = \sum_i \Phi(x_i) f(x_i)$ where f is the distribution of X.

(Proof is given for the case X is discrete and finite.)

Suppose that X takes on the values $x_1, \ldots, x_n$ and that $\Phi(x_i)$ takes on the values $y_1, \ldots, y_m$ as i runs from 1 to n. Then clearly the possible values of $Y = \Phi(X)$ are $y_1, \ldots, y_m$ and the distribution g of Y is given by

$$g(y_j) = \sum_{\{i\,:\,\Phi(x_i)=y_j\}} f(x_i)$$

Therefore

$$E(Y) = \sum_{j=1}^{m} y_j\, g(y_j) = \sum_{j=1}^{m} y_j \sum_{\{i\,:\,\Phi(x_i)=y_j\}} f(x_i)$$

$$= \sum_{i=1}^{n} f(x_i) \sum_{\{j\,:\,\Phi(x_i)=y_j\}} y_j = \sum_{i=1}^{n} f(x_i)\,\Phi(x_i)$$

which proves the theorem.

5.14. Prove Theorem 5.1: Let X be a random variable and k a real number. Then (i) $E(kX) = k\,E(X)$ and (ii) $E(X+k) = E(X) + k$.

(Proof is given for the general discrete case with the assumption that $E(X)$ exists.)

(i) Now $kX = \Phi(X)$ where $\Phi(x) = kx$. Therefore by Theorem 5.8 (Problem 5.13),

$$E(kX) = \sum_i kx_i\, f(x_i) = k \sum_i x_i\, f(x_i) = k\,E(X)$$

(ii) Here $X + k = \Phi(X)$ where $\Phi(x) = x + k$. Therefore

$$E(X+k) = \sum_i (x_i + k)\, f(x_i) = \sum_i x_i\, f(x_i) + \sum_i k\, f(x_i) = E(X) + k$$

5.15. Prove Theorem 5.2: Let X and Y be random variables on the same sample space S. Then $E(X+Y) = E(X) + E(Y)$.

(Proof is given for the general discrete case with the assumption that $E(X)$ and $E(Y)$ both exist.)

Now $X + Y = \Phi(X, Y)$ where $\Phi(x, y) = x + y$. Therefore by Theorem 5.9,

$$E(X+Y) = \sum_i \sum_j (x_i + y_j)\, h(x_i, y_j) = \sum_i \sum_j x_i\, h(x_i, y_j) + \sum_i \sum_j y_j\, h(x_i, y_j)$$

Applying Problem 5.12, we get

$$E(X+Y) = \sum_i x_i\, f(x_i) + \sum_j y_j\, g(y_j) = E(X) + E(Y)$$

5.16. Prove Corollary 5.3: Let $X_1, X_2, \ldots, X_n$ be random variables on S. Then

$$E(X_1 + \cdots + X_n) = E(X_1) + \cdots + E(X_n)$$

(Proof is given for the general discrete case with the assumption that $E(X_1), \ldots, E(X_n)$ all exist.)

We prove this by induction on n. The case $n = 1$ is trivial and the case $n = 2$ is just Theorem 5.2 (Problem 5.15). For the case $n > 2$ we apply the case $n = 2$ to obtain

$$E(X_1 + \cdots + X_{n-1} + X_n) = E(X_1 + \cdots + X_{n-1}) + E(X_n)$$

and by the inductive hypothesis this becomes $E(X_1) + \cdots + E(X_{n-1}) + E(X_n)$.

5.17. Prove Theorem 5.5: (i) $\mathrm{Var}(X+k) = \mathrm{Var}(X)$ and (ii) $\mathrm{Var}(kX) = k^2\,\mathrm{Var}(X)$. Hence $\sigma_{X+k} = \sigma_X$ and $\sigma_{kX} = |k|\,\sigma_X$.

By Theorem 5.1, $\mu_{X+k} = \mu_X + k$ and $\mu_{kX} = k\mu_X$. Also $\sum x_i\, f(x_i) = \mu_X$ and $\sum f(x_i) = 1$. Hence

$$\begin{aligned}\text{Var}(X+k) &= \sum (x_i+k)^2 f(x_i) - \mu_{X+k}^2 \\ &= \sum x_i^2 f(x_i) + 2k\sum x_i f(x_i) + k^2 \sum f(x_i) - (\mu_X + k)^2 \\ &= \sum x_i^2 f(x_i) + 2k\mu_X + k^2 - (\mu_X^2 + 2k\mu_X + k^2) \\ &= \sum x_i f(x_i) - \mu_X^2 = \text{Var}(X)\end{aligned}$$

and

$$\begin{aligned}\text{Var}(kX) &= \sum (kx_i)^2 f(x_i) - \mu_{kX}^2 = k^2 \sum x_i^2 f(x_i) - (k\mu_X)^2 \\ &= k^2 \sum x_i^2 f(x_i) - k^2\mu_X^2 = k^2\left(\sum x_i^2 f(x_i) - \mu_X^2\right) = k^2 \text{Var}(X)\end{aligned}$$

5.18. Show that

$$\text{Cov}(X,Y) = \sum_{i,j} (x_i - \mu_X)(y_j - \mu_Y)\, h(x_i, y_j) = \sum_{i,j} x_i y_j\, h(x_i, y_j) - \mu_X \mu_Y$$

(Proof is given for the case when X and Y are discrete and finite.)

Since

$$\sum_{i,j} y_j\, h(x_i, y_j) = \sum_j y_j\, g(y_j) = \mu_Y, \quad \sum_{i,j} x_i\, h(x_i, y_j) = \sum_i x_i f(x_i) = \mu_X \quad \text{and} \quad \sum_{i,j} h(x_i, y_j) = 1$$

we obtain

$$\begin{aligned}&\sum_{i,j} (x_i - \mu_X)(y_j - \mu_Y)\, h(x_i, y_j) \\ &\quad = \sum_{i,j} (x_i y_j - \mu_X y_j - \mu_Y x_i + \mu_X \mu_Y)\, h(x_i, y_j) \\ &\quad = \sum_{i,j} x_i y_j\, h(x_i, y_j) - \mu_X \sum_{i,j} y_j\, h(x_i, y_j) - \mu_Y \sum_{i,j} x_i\, h(x_i, y_j) + \mu_X\mu_Y \sum_{i,j} h(x_i, y_j) \\ &\quad = \sum_{i,j} x_i y_j\, h(x_i, y_j) - \mu_X\mu_Y - \mu_X\mu_Y + \mu_X\mu_Y \\ &\quad = \sum_{i,j} x_i y_j\, h(x_i, y_j) - \mu_X\mu_Y\end{aligned}$$

5.19. Prove Theorem 5.6: Let X and Y be independent random variables. Then (i) $E(XY) = E(X)\,E(Y)$, (ii) $\text{Var}(X+Y) = \text{Var}(X) + \text{Var}(Y)$, (iii) $\text{Cov}(X,Y) = 0$.

(Proof is given for the case when X and Y are discrete and finite.)

Since X and Y are independent, $h(x_i, y_j) = f(x_i)\, g(y_j)$. Thus

$$\begin{aligned}E(XY) &= \sum_{i,j} x_i y_j\, h(x_i, y_j) = \sum_{i,j} x_i y_j\, f(x_i)\, g(y_j) \\ &= \sum_i x_i f(x_i) \sum_j y_j\, g(y_j) = E(X)\,E(Y)\end{aligned}$$

and

$$\text{Cov}(X,Y) = E(XY) - \mu_X\mu_Y = E(X)\,E(Y) - \mu_X\mu_Y = 0$$

In order to prove (ii) we also need

$$\mu_{X+Y} = \mu_X + \mu_Y, \quad \sum_{i,j} x_i^2\, h(x_i, y_j) = \sum_i x_i^2 f(x_i), \quad \sum_{i,j} y_j^2\, h(x_i, y_j) = \sum_j y_j^2\, g(y_j)$$

Hence

$$\begin{aligned}\text{Var}(X+Y) &= \sum_{i,j} (x_i + y_j)^2\, h(x_i, y_j) - \mu_{X+Y}^2 \\ &= \sum_{i,j} x_i^2\, h(x_i, y_j) + 2\sum_{i,j} x_i y_j\, h(x_i, y_j) + \sum_{i,j} y_j^2\, h(x_i, y_j) - (\mu_X + \mu_Y)^2 \\ &= \sum_i x_i^2 f(x_i) + 2\sum_i x_i f(x_i) \sum_j y_j\, g(y_j) + \sum_j y_j^2\, g(y_j) - \mu_X^2 - 2\mu_X\mu_Y - \mu_Y^2 \\ &= \sum_i x_i^2 f(x_i) - \mu_X^2 + \sum_j y_j^2\, g(y_j) - \mu_Y^2 = \text{Var}(X) + \text{Var}(Y)\end{aligned}$$

5.20. Prove Theorem 5.7: Let $X_1, X_2, \ldots, X_n$ be independent random variables. Then

$$\operatorname{Var}(X_1 + \cdots + X_n) = \operatorname{Var}(X_1) + \cdots + \operatorname{Var}(X_n)$$

(Proof is given for the case when $X_1, \ldots, X_n$ are all discrete and finite.)

We take for granted the analogs of Problem 5.12 and Theorem 5.9 for n random variables. Then

$$\begin{aligned}
\operatorname{Var}(X_1 + \cdots + X_n) &= E((X_1 + \cdots + X_n - \mu_{X_1 + \cdots + X_n})^2) \\
&= \sum (x_1 + \cdots + x_n - \mu_{X_1 + \cdots + X_n})^2 \, h(x_1, \ldots, x_n) \\
&= \sum (x_1 + \cdots + x_n - \mu_{X_1} - \cdots - \mu_{X_n})^2 \, h(x_1, \ldots, x_n) \\
&= \sum \left\{ \sum_i \sum_j x_i x_j + \sum_i \sum_j \mu_{X_i}\mu_{X_j} - 2 \sum_i \sum_j \mu_{X_i} x_j \right\} h(x_1, \ldots, x_n)
\end{aligned}$$

where h is the joint distribution of $X_1, \ldots, X_n$, and $\mu_{X_1 + \cdots + X_n} = \mu_{X_1} + \cdots + \mu_{X_n}$ (Corollary 5.3). Since the X_i are pairwise independent, $\sum x_i x_j \, h(x_1, \ldots, x_n) = \mu_{X_i}\mu_{X_j}$ for $i \neq j$. Hence

$$\begin{aligned}
\operatorname{Var}(X_1 + \cdots + X_n) &= \sum_{i \neq j} \mu_{X_i}\mu_{X_j} + \sum_{i=1}^{n} E(X_i^2) + \sum_i \sum_j \mu_{X_i}\mu_{X_j} - 2 \sum_i \sum_j \mu_{X_i}\mu_{X_j} \\
&= \sum_{i=1}^{n} E(X_i^2) - \sum_{i=1}^{n} (\mu_{X_i})^2 = \sum_{i=1}^{n} \operatorname{Var}(X_i)
\end{aligned}$$

as required.

MISCELLANEOUS PROBLEMS

5.21. Let X be a continuous random variable with distribution

$$f(x) = \begin{cases} \frac{1}{6}x + k & \text{if } 0 \leq x \leq 3 \\ 0 & \text{elsewhere} \end{cases}$$

(i) Evaluate k. (ii) Find $P(1 \leq X \leq 2)$.

(i) The graph of f is drawn below. Since f is a continuous probability function, the shaded region A must have area 1. Note A forms a trapezoid with parallel bases of lengths k and $k + \frac{1}{2}$, and altitude 3. Hence the area of $A = \frac{1}{2}(k + k + \frac{1}{2}) \cdot 3 = 1$ or $k = \frac{1}{12}$.

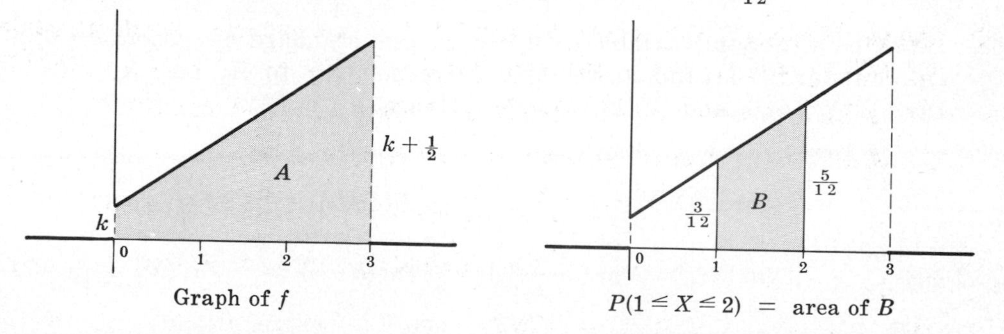

Graph of f $\qquad$ $P(1 \leq X \leq 2)$ = area of B

(ii) $P(1 \leq X \leq 2)$ is equal to the area of B which is under the graph of f and between $x = 1$ and $x = 2$ as shown above on the right. Note $f(1) = \frac{1}{6} + \frac{1}{12} = \frac{3}{12}$, $f(2) = \frac{1}{3} + \frac{1}{12} = \frac{5}{12}$. Hence $P(1 \leq X \leq 2)$ = area of B = $\frac{1}{2}(\frac{3}{12} + \frac{5}{12}) \cdot 1 = \frac{1}{3}$.

5.22. Let X be a continuous random variable whose distribution f is constant on an interval, say $I = \{a \leq x \leq b\}$, and 0 elsewhere:

$$f(x) = \begin{cases} k & \text{if } a \leq x \leq b \\ 0 & \text{elsewhere} \end{cases}$$

(Such a random variable is said to be *uniformly distributed* on I.) (i) Determine k. (ii) Find the mean μ of X. (iii) Determine the cumulative distribution function F of X.

(i) The graph of f appears on the right. The region A must have area 1; hence

$$k(b-a) = 1 \quad \text{or} \quad k = \frac{1}{b-a}$$

(ii) If we view probability as weight or mass, and the mean as the center of gravity, then it is intuitively clear that

$$\mu = \frac{a+b}{2}$$

Graph of f

the point midway between a and b. We verify this mathematically using calculus:

$$\mu = E(X) = \int_{\mathbf{R}} x\,f(x)\,dx = \int_a^b \frac{x}{b-a}\,dx = \left[\frac{x^2}{2(b-a)}\right]_a^b$$

$$= \frac{b^2}{2(b-a)} - \frac{a^2}{2(b-a)} = \frac{a+b}{2}$$

(iii) Recall that the cumulative distribution function F is defined by $F(k) = P(X \leq k)$. Hence $F(k)$ gives the area under the graph of f to the left of $x = k$. Since X is uniformly distributed on the interval $I = \{a \leq x \leq b\}$, it is intuitive that the graph of F should be as shown on the right, i.e. $F \equiv 0$ before the point a, $F \equiv 1$ after the point b, and F is linear between a and b. We verify this mathematically using calculus:

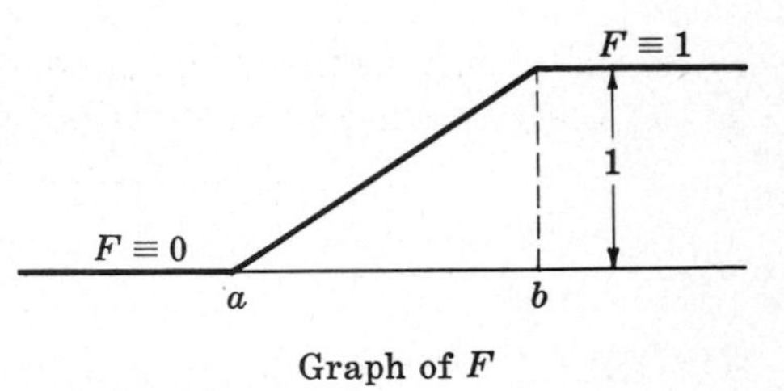

Graph of F

(a) for $x < a$,

$$F(x) = \int_{-\infty}^{x} f(t)\,dt = \int_{-\infty}^{x} 0\,dt = 0$$

(b) for $a \leq x \leq b$,

$$F(x) = \int_{-\infty}^{x} f(t)\,dt = \int_a^x \frac{1}{b-a}\,dt = \left[\frac{t}{b-a}\right]_a^x = \frac{x-a}{b-a}$$

(c) for $x > b$, $F(x) = P(X \leq x) \geq P(X \leq b) = F(b) = 1$ and also $1 \geq P(X \leq x) = F(x)$; hence $F(x) = 1$.

5.23. Let X be a random variable with mean μ and standard deviation $\sigma > 0$; and let X^* be the standardized random variable corresponding to X, i.e. $X^* = (X - \mu)/\sigma$. Show that $E(X^*) = 0$ and $\text{Var}(X^*) = 1$. (Hence $\sigma_{X^*} = 1$.)

By Theorem 5.1 and Theorem 5.5,

$$E(X^*) = E\left(\frac{X-\mu}{\sigma}\right) = \frac{1}{\sigma}E(X-\mu) = \frac{1}{\sigma}(E(X)-\mu) = 0$$

and

$$\text{Var}(X^*) = \text{Var}\left(\frac{X-\mu}{\sigma}\right) = \frac{1}{\sigma^2}\text{Var}(X-\mu) = \frac{1}{\sigma^2}\text{Var}(X) = \frac{\sigma^2}{\sigma^2} = 1$$

5.24. Let X be a random variable with distribution f. The rth *moment* M_r of X is defined by

$$M_r = E(X^r) = \sum x_i^r\,f(x_i)$$

Find the first five moments of X if X has the following distribution:

x_i	-2	1	3
$f(x_i)$	$\frac{1}{2}$	$\frac{1}{4}$	$\frac{1}{4}$

(Note that M_1 is the mean of X, and M_2 is used in computing the variance and standard deviation of X.)

$$M_1 = \sum x_i f(x_i) = -2 \cdot \tfrac{1}{2} + 1 \cdot \tfrac{1}{4} + 3 \cdot \tfrac{1}{4} = 0,$$

$$M_2 = \sum x_i^2 f(x_i) = 4 \cdot \tfrac{1}{2} + 1 \cdot \tfrac{1}{4} + 9 \cdot \tfrac{1}{4} = 4.5,$$

$$M_3 = \sum x_i^3 f(x_i) = -8 \cdot \tfrac{1}{2} + 1 \cdot \tfrac{1}{4} + 27 \cdot \tfrac{1}{4} = 3,$$

$$M_4 = \sum x_i^4 f(x_i) = 16 \cdot \tfrac{1}{2} + 1 \cdot \tfrac{1}{4} + 81 \cdot \tfrac{1}{4} = 28.5,$$

$$M_5 = \sum x_i^5 f(x_i) = -32 \cdot \tfrac{1}{2} + 1 \cdot \tfrac{1}{4} + 243 \cdot \tfrac{1}{4} = 45.$$

5.25. Let h be the joint distribution of the random variables X and Y. (i) Show that the distribution f of the sum $Z = X + Y$ can be obtained by summing the probabilities along the diagonal lines $x + y = z_k$, i.e.

$$f(z_k) = \sum_{z_k = x_i + y_j} h(x_i, y_j) = \sum_{x_i} h(x_i, z_k - x_i)$$

(ii) Apply (i) to obtain the distribution f of the sum $Z = X + Y$ where X and Y have the following joint distribution:

X \ Y	−2	−1	0	1	2	3	Sum
0	.05	.05	.10	0	.05	.05	.30
1	.10	.05	.05	.10	0	.05	.35
2	.03	.12	.07	.06	.03	.04	.35
Sum	.18	.22	.22	.16	.08	.14	

(i) The events $\{X = x_i, Y = y_j : x_i + y_j = z_k\}$ are disjoint; hence

$$f(z_k) = P(Z = z_k) = \sum_{x_i + y_j = z_k} P(X = x_i, Y = y_j)$$

$$= \sum_{x_i + y_j = z_k} h(x_i, y_j) = \sum_{x_i} h(x_i, z_k - x_i)$$

(ii)

X \ Y	−2	−1	0	1	2	3
0	.05	.05	.10	0	.05	.05
1	.10	.05	.05	.10	0	.05
2	.03	.12	.07	.06	.03	.04

Adding along the diagonal lines in the above table, we obtain

$f(-2) = .05$

$f(-1) = .05 + .10 = .15$

$f(0) = .10 + .05 + .03 = .18$

$f(1) = 0 + .05 + .12 = .17$

$f(2) = .05 + .10 + .07 = .22$

$f(3) = .05 + 0 + .06 = .11$

$f(4) = .05 + .03 = .08$

$f(5) = .04$

In other words, the distribution of $Z = X + Y$ is as follows:

z_i	−2	−1	0	1	2	3	4	5
$f(z_i)$	.05	.15	.18	.17	.22	.11	.08	.04

Supplementary Problems

RANDOM VARIABLES

5.26. Find the mean μ, variance σ^2 and standard deviation σ of each distribution:

(i)

x_i	2	3	8
$f(x_i)$	$\frac{1}{4}$	$\frac{1}{2}$	$\frac{1}{4}$

(ii)

x_i	-2	-1	7
$f(x_i)$	$\frac{1}{3}$	$\frac{1}{2}$	$\frac{1}{6}$

(iii)

x_i	-1	0	1	2	3
$f(x_i)$	.3	.1	.1	.3	.2

5.27. A pair of fair dice is thrown. Let X be the random variable which denotes the minimum of the two numbers which appear. Find the distribution, mean, variance and standard deviation of X.

5.28. A fair coin is tossed four times. Let X denote the number of heads occurring. Find the distribution, mean, variance and standard deviation of X.

5.29. A fair coin is tossed four times. Let Y denote the longest string of heads occurring. Find the distribution, mean, variance and standard deviation of Y.

5.30. Find the mean μ, variance σ^2 and standard deviation σ of the two-point distribution

x_i	a	b
$f(x_i)$	p	q

where $p+q=1$.

5.31. Two cards are selected at random from a box which contains five cards numbered 1, 1, 2, 2 and 3. Let X denote the sum and Y the maximum of the two numbers drawn. Find the distribution, mean, variance and standard deviation of (i) X, (ii) Y, (iii) $X+Y$, (iv) XY.

EXPECTATION

5.32. A fair coin is tossed until a head or four tails occur. Find the expected number of tosses of the coin.

5.33. A coin weighted so that $P(\text{H}) = \frac{1}{3}$ and $P(\text{T}) = \frac{2}{3}$ is tossed until a head or five tails occur. Find the expected number of tosses of the coin.

5.34. A box contains 8 items of which 2 are defective. A man selects 3 items from the box. Find the expected number of defective items he has drawn.

5.35. A box contains 10 transistors of which 2 are defective. A transistor is selected from the box and tested until a nondefective one is chosen. Find the expected number of transistors to be chosen.

5.36. Solve the preceding problem in the case that 3 of the 10 items are defective.

5.37. The probability of team A winning any game is $\frac{1}{2}$. A plays team B in a tournament. The first team to win 2 games in a row or a total of three games wins the tournament. Find the expected number of games in the tournament.

5.38. A player tosses three fair coins. He wins \$5 if 3 heads occur, \$3 if 2 heads occur, and \$1 if only 1 head occurs. On the other hand, he loses \$15 if 3 tails occur. Find the value of the game to the player.

5.39. A player tosses three fair coins. He wins \$8 if 3 heads occur, \$3 if 2 heads occur, and \$1 if only 1 head occurs. If the game is to be fair, how much should he lose if no heads occur?

5.40. A player tosses three fair coins. He wins \$10 if 3 heads occur, \$5 if 2 heads occur, \$3 if 1 head occurs and \$2 if no heads occur. If the game is to be fair, how much should he pay to play the game?

JOINT DISTRIBUTION, INDEPENDENT RANDOM VARIABLES

5.41. Consider the following joint distribution of X and Y:

$X \backslash Y$	-4	2	7	Sum
1	$\frac{1}{8}$	$\frac{1}{4}$	$\frac{1}{8}$	$\frac{1}{2}$
5	$\frac{1}{4}$	$\frac{1}{8}$	$\frac{1}{8}$	$\frac{1}{2}$
Sum	$\frac{3}{8}$	$\frac{3}{8}$	$\frac{1}{4}$	

Find (i) $E(X)$ and $E(Y)$, (ii) $\text{Cov}(X, Y)$, (iii) σ_X, σ_Y and $\rho(X, Y)$.

5.42. Consider the following joint distribution of X and Y:

$X \backslash Y$	-2	-1	4	5	Sum
1	.1	.2	0	.3	.6
2	.2	.1	.1	0	.4
Sum	.3	.3	.1	.3	

Find (i) $E(X)$ and $E(Y)$, (ii) $\text{Cov}(X, Y)$, (iii) σ_X, σ_Y and $\rho(X, Y)$.

5.43. Suppose X and Y are independent random variables with the following respective distributions:

x_i	1	2
$f(x_i)$	.7	.3

y_j	-2	5	8
$g(y_j)$	.3	.5	.2

Find the joint distribution of X and Y, and verify that $\text{Cov}(X, Y) = 0$.

5.44. A fair coin is tossed four times. Let X denote the number of heads occurring and let Y denote the longest string of heads occurring (see Problems 5.28 and 5.29). (i) Determine the joint distribution of X and Y. (ii) Find $\text{Cov}(X, Y)$ and $\rho(X, Y)$.

5.45. Two cards are selected at random from a box which contains five cards numbered 1, 1, 2, 2 and 3. Let X denote the sum and Y the maximum of the two numbers drawn (see Problem 5.31). (i) Determine the joint distribution of X and Y. (ii) Find $\text{Cov}(X, Y)$ and $\rho(X, Y)$.

MISCELLANEOUS PROBLEMS

5.46. Let X be a continuous random variable with distribution

$$f(x) = \begin{cases} \frac{1}{8} & \text{if } 0 \le x \le 8 \\ 0 & \text{elsewhere} \end{cases}$$

(i) Find: $P(2 \le X \le 5)$, $P(3 \le X \le 7)$ and $P(X \ge 6)$.

(ii) Determine and plot the graph of the cumulative distribution function F of X.

5.47. Let X be a continuous random variable with distribution

$$f(x) = \begin{cases} kx & \text{if } 0 \le x \le 5 \\ 0 & \text{elsewhere} \end{cases}$$

(i) Evaluate k. (ii) Find $P(1 \le X \le 3)$, $P(2 \le X \le 4)$ and $P(X \le 3)$.

5.48. Plot the graph of the cumulative distribution function F of the random variable X with distribution

x_i	-3	2	6
$f(x_i)$	$\frac{1}{4}$	$\frac{1}{2}$	$\frac{1}{4}$

5.49. Show that $\sigma_X = 0$ if and only if X is a *constant function*, i.e. $X(s) = k$ for every $s \in S$, or simply $X = k$.

5.50. If $\sigma_X \neq 0$, show that $\rho(X, X) = 1$ and $\rho(X, -X) = -1$.

5.51. Prove Theorem 5.9: Let X, Y and Z be random variables on S with $Z = \Phi(X, Y)$. Then

$$E(Z) = \sum_{i,j} \Phi(x_i, y_j)\, h(x_i, y_j)$$

where h is the joint distribution of X and Y.

Answers to Supplementary Problems

5.26. (i) $\mu = 4$, $\sigma^2 = 5.5$, $\sigma = 2.3$; (ii) $\mu = 0$, $\sigma^2 = 10$, $\sigma = 3.2$; (iii) $\mu = 1$, $\sigma^2 = 2.4$, $\sigma = 1.5$.

5.27.

x_i	1	2	3	4	5	6
$f(x_i)$	$\frac{11}{36}$	$\frac{9}{36}$	$\frac{7}{36}$	$\frac{5}{36}$	$\frac{3}{36}$	$\frac{1}{36}$

$E(X) = 2.5$, $\text{Var}(X) = 2.1$, $\sigma_X = 1.4$

5.28.

x_i	0	1	2	3	4
$f(x_i)$	$\frac{1}{16}$	$\frac{4}{16}$	$\frac{6}{16}$	$\frac{4}{16}$	$\frac{1}{16}$

$E(X) = 2$, $\text{Var}(X) = 1$, $\sigma_X = 1$

5.29.

y_j	0	1	2	3	4
$g(y_j)$	$\frac{1}{16}$	$\frac{7}{16}$	$\frac{5}{16}$	$\frac{2}{16}$	$\frac{1}{16}$

$E(Y) = 1.7$, $\text{Var}(Y) = 0.9$, $\sigma_Y = 0.95$

5.30. $\mu = ap + bq$, $\sigma^2 = pq(a-b)^2$, $\sigma = |a-b|\sqrt{pq}$

5.31. (i)

x_i	2	3	4	5
$f(x_i)$	.1	.4	.3	.2

$E(X) = 3.6$, $\text{Var}(X) = .84$, $\sigma_X = .9$

(ii)

y_j	1	2	3
$g(y_j)$	.1	.5	.4

$E(Y) = 2.3$, $\text{Var}(Y) = .41$, $\sigma_Y = .64$

(iii)

z_k	3	5	6	7	8
$p(z_k)$	.1	.4	.1	.2	.2

$E(X+Y) = 5.9$, $\text{Var}(X+Y) = 2.3$, $\sigma_{X+Y} = 1.5$

(iv)

w_k	2	6	8	12	15
$s(w_k)$	.1	.4	.1	.2	.2

$E(XY) = 8.8$, $\text{Var}(XY) = 17.6$, $\sigma_{XY} = 4.2$

5.32. 15/8

5.33. 211/81

5.34. 3/4

5.35. 11/9

5.36. 11/8

5.37. 23/8

5.38. 25¢ in favor of the player

5.39. $20

5.40. $4.50

5.41. (i) $E(X) = 3$, $E(Y) = 1$; (ii) $\text{Cov}(X,Y) = 1.5$; (iii) $\sigma_X = 2$, $\sigma_Y = 4.3$, $\rho(X,Y) = .17$

5.42. (i) $E(X) = 1.4$, $E(Y) = 1$; (ii) $\text{Cov}(X,Y) = -.5$; (iii) $\sigma_X = .49$, $\sigma_Y = 3.1$, $\rho(X,Y) = -.3$

5.43.

X \ Y	−2	5	8	Sum
1	.21	.35	.14	.7
2	.09	.15	.06	.3
Sum	.3	.5	.2	

5.44. (i)

X \ Y	0	1	2	3	4	Sum
0	$\frac{1}{16}$	0	0	0	0	$\frac{1}{16}$
1	0	$\frac{4}{16}$	0	0	0	$\frac{4}{16}$
2	0	$\frac{3}{16}$	$\frac{3}{16}$	0	0	$\frac{6}{16}$
3	0	0	$\frac{2}{16}$	$\frac{2}{16}$	0	$\frac{4}{16}$
4	0	0	0	0	$\frac{1}{16}$	$\frac{1}{16}$
Sum	$\frac{1}{16}$	$\frac{7}{16}$	$\frac{5}{16}$	$\frac{2}{16}$	$\frac{1}{16}$	

(ii) $\text{Cov}(X,Y) = .85$, $\rho(X,Y) = .89$

5.45. (i)

X \ Y	1	2	3	Sum
2	.1	0	0	.1
3	0	.4	0	.4
4	0	.1	.2	.3
5	0	0	.2	.2
Sum	.1	.5	.4	

(ii) $\text{Cov}(X, Y) = .52, \quad \rho(X, Y) = .9$

5.46. (i) $P(2 \leq X \leq 5) = \frac{3}{8}$, $P(3 \leq X \leq 7) = \frac{1}{2}$, $P(X \geq 6) = \frac{1}{4}$

(ii) $$F(x) = \begin{cases} 0 & \text{if } x < 0 \\ \frac{1}{8}x & \text{if } 0 \leq x \leq 8 \\ 1 & \text{if } x > 8 \end{cases}$$

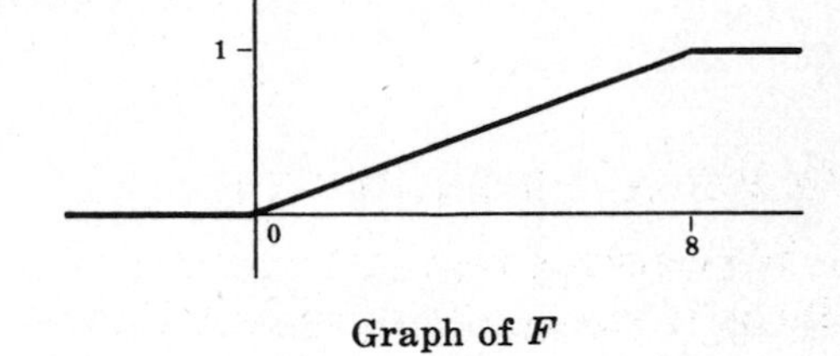

Graph of F

5.47. (i) $k = \frac{2}{25}$, (ii) $P(1 \leq X \leq 3) = \frac{8}{25}$, $P(2 \leq X \leq 4) = \frac{12}{25}$, $P(X \leq 3) = \frac{9}{25}$

5.48.

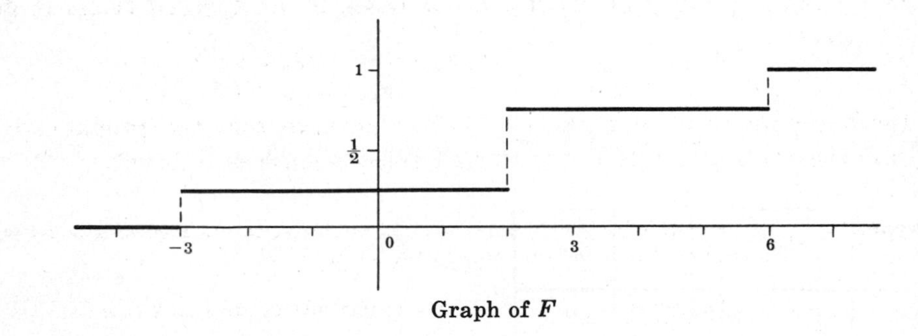

Graph of F

Chapter 6

Binomial, Normal and Poisson Distributions

BINOMIAL DISTRIBUTION

We consider repeated and independent trials of an experiment with two outcomes; we call one of the outcomes *success* and the other outcome *failure*. Let p be the probability of success, so that $q = 1 - p$ is the probability of failure. If we are interested in the number of successes and not in the order in which they occur, then the following theorem applies.

Theorem 6.1: The probability of exactly k successes in n repeated trials is denoted and given by

$$b(k; n, p) = \binom{n}{k} p^k q^{n-k}$$

Here $\binom{n}{k}$ is the binomial coefficient (see page 19). Observe that the probability of no successes is q^n, and therefore the probability of at least one success is $1 - q^n$.

Example 6.1: A fair coin is tossed 6 times or, equivalently, six fair coins are tossed; call heads a success. Then $n = 6$ and $p = q = \frac{1}{2}$.

(i) The probability that exactly two heads occur (i.e. $k = 2$) is

$$b(2; 6, \tfrac{1}{2}) = \binom{6}{2} (\tfrac{1}{2})^2 (\tfrac{1}{2})^4 = \tfrac{15}{64}$$

(ii) The probability of getting at least four heads (i.e. $k = 4$, 5 or 6) is

$$\begin{aligned} b(4; 6, \tfrac{1}{2}) + b(5; 6, \tfrac{1}{2}) + b(6; 6, \tfrac{1}{2}) &= \binom{6}{4} (\tfrac{1}{2})^4 (\tfrac{1}{2})^2 + \binom{6}{5} (\tfrac{1}{2})^5 (\tfrac{1}{2}) + \binom{6}{6} (\tfrac{1}{2})^6 \\ &= \tfrac{15}{64} + \tfrac{6}{64} + \tfrac{1}{64} = \tfrac{11}{32} \end{aligned}$$

(iii) The probability of no heads (i.e. all failures) is $q^6 = (\frac{1}{2})^6 = \frac{1}{64}$, and so the probability of at least one head is $1 - q^6 = 1 - \frac{1}{64} = \frac{63}{64}$.

Example 6.2: A fair die is tossed 7 times; call a toss a success if a 5 or a 6 appears. Then $n = 7$, $p = P(\{5, 6\}) = \frac{1}{3}$ and $q = 1 - p = \frac{2}{3}$.

(i) The probability that a 5 or a 6 occurs exactly 3 times (i.e. $k = 3$) is

$$b(3; 7, \tfrac{1}{3}) = \binom{7}{3} (\tfrac{1}{3})^3 (\tfrac{2}{3})^4 = \tfrac{560}{2187}$$

(ii) The probability that a 5 or a 6 never occurs (i.e. all failures) is $q^7 = (\frac{2}{3})^7 = \frac{128}{2187}$; hence the probability that a 5 or a 6 occurs at least once is $1 - q^7 = \frac{2059}{2187}$.

If we regard n and p as constant, then the above function $P(k) = b(k; n, p)$ is a discrete probability distribution:

k	0	1	2	$\cdots$	n
$P(k)$	q^n	$\binom{n}{1} q^{n-1} p$	$\binom{n}{2} q^{n-2} p^2$	$\cdots$	p^n

It is called the *binomial distribution* since for $k = 0, 1, 2, \ldots, n$ it corresponds to the successive terms of the binomial expansion

$$(q+p)^n = q^n + \binom{n}{1} q^{n-1} p + \binom{n}{2} q^{n-2} p^2 + \cdots + p^n$$

This distribution is also called the Bernoulli distribution, and independent trials with two outcomes are called Bernoulli trials.

Properties of this distribution follow:

Theorem 6.2:

Binomial distribution	
Mean	$\mu = np$
Variance	$\sigma^2 = npq$
Standard deviation	$\sigma = \sqrt{npq}$

Example 6.3: A fair die is tossed 180 times. The expected number of sixes is $\mu = np = 180 \cdot \frac{1}{6} = 30$. The standard deviation is $\sigma = \sqrt{npq} = \sqrt{180 \cdot \frac{1}{6} \cdot \frac{5}{6}} = 5$.

NORMAL DISTRIBUTION

The *normal* (or: *Gaussian*) *distribution* or *curve* is defined as follows:

$$f(x) = \frac{1}{\sigma\sqrt{2\pi}} e^{-\frac{1}{2}(x-\mu)^2/\sigma^2}$$

where μ and $\sigma > 0$ are arbitrary constants. This function is certainly one of the most important examples of a continuous probability distribution. The two diagrams below show the changes in f as μ varies and as σ varies. In particular, observe that these bell-shaped curves are symmetric about $x = \mu$.

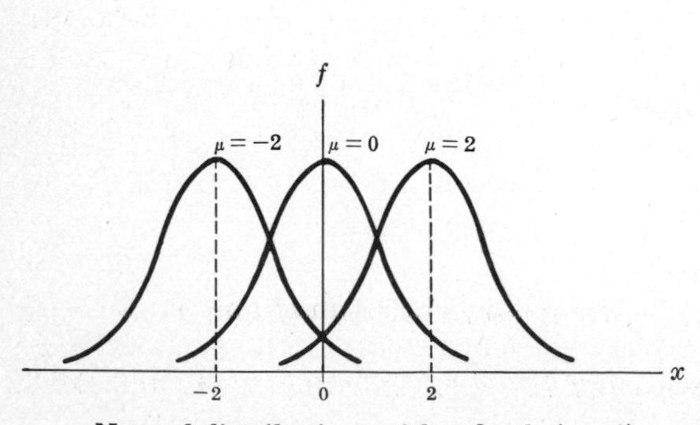

Normal distributions with σ fixed ($\sigma = 1$)

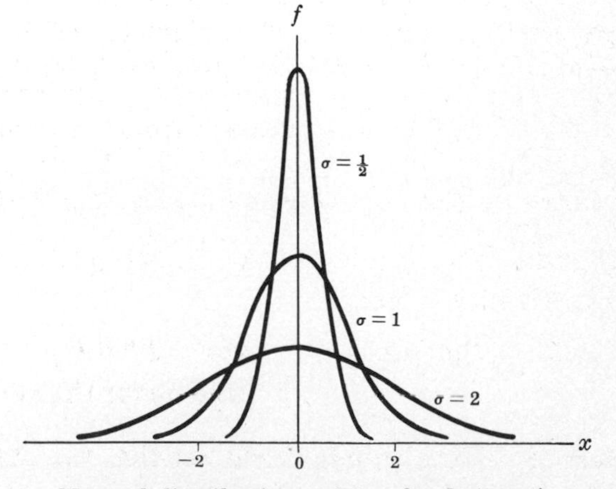

Normal distributions with μ fixed ($\mu = 0$)

Properties of the normal distribution follow:

Theorem 6.3:

Normal distribution	
Mean	μ
Variance	σ^2
Standard deviation	σ

We denote the above normal distribution with mean μ and variance σ^2 by

$$N(\mu, \sigma^2)$$

If we make the substitution $t = (x - \mu)/\sigma$ in the above formula for $N(\mu, \sigma^2)$ we obtain the *standard normal distribution* or *curve*

$$\phi(t) \quad = \quad \frac{1}{\sqrt{2\pi}} e^{-\frac{1}{2}t^2}$$

which has mean $\mu = 0$ and variance $\sigma^2 = 1$. The graph of this distribution appears below. We note that for $-1 \leq t \leq 1$ we obtain 68.2% of the area under the curve, and for $-2 \leq t \leq 2$ we obtain 95.4% of the area under the curve.

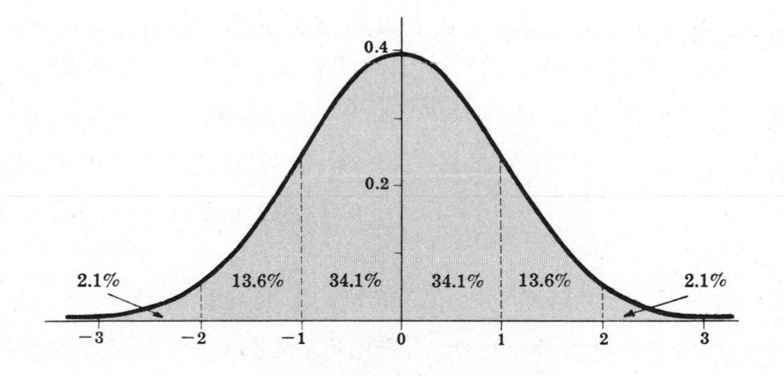

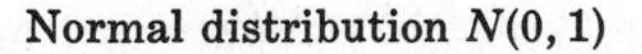
Normal distribution $N(0,1)$

A table on page 111 gives the area under the standard normal curve between $t = 0$ and any positive value of t. The symmetry of the curve about $t = 0$ permits us to obtain the area between any two values of t (see Problem 6.14).

Now let X be a continuous random variable with a normal distribution; we frequently say that X is *normally distributed*. We compute the probability that X lies between a and b, denoted by $P(a \leq X \leq b)$, as follows. First we change a and b into *standard units*

$$a' \;=\; (a - \mu)/\sigma \quad \text{and} \quad b' \;=\; (b - \mu)/\sigma$$

respectively. Then

$$\begin{aligned} P(a \leq X \leq b) \;&=\; P(a' \leq X^* \leq b') \\ &=\; \text{area under the standard normal curve between } a' \text{ and } b' \end{aligned}$$

Here X^* is the standardized random variable (see page 79) corresponding to X, and hence X^* has the standard normal distribution $N(0,1)$.

NORMAL APPROXIMATION TO THE BINOMIAL DISTRIBUTION. CENTRAL LIMIT THEOREM

The binomial distribution $P(k) = b(k; n, p)$ is closely approximated by the normal distribution providing n is large and neither p nor q is close to zero. This property is indicated in the following diagram where we have chosen the binomial distribution corresponding to $n = 8$ and $p = q = \frac{1}{2}$.

k	0	1	2	3	4	5	6	7	8
$P(k)$	$\frac{1}{256}$	$\frac{8}{256}$	$\frac{28}{256}$	$\frac{56}{256}$	$\frac{70}{256}$	$\frac{56}{256}$	$\frac{28}{256}$	$\frac{8}{256}$	$\frac{1}{256}$

Binomial distribution with $n = 8$ and $p = q = \frac{1}{2}$

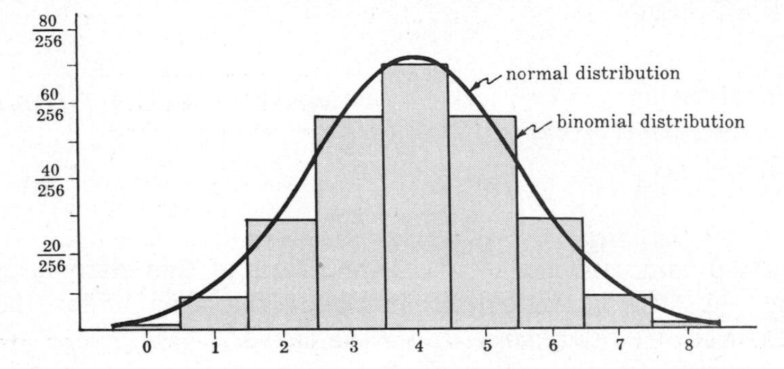

Comparison of the binomial and normal distributions

The above property of the normal distribution is generalized in the Central Limit Theorem which follows. The proof of this theorem lies beyond the scope of this text.

Central Limit Theorem 6.4: Let $X_1, X_2, \ldots$ be a sequence of independent random variables with the same distribution with mean μ and variance σ^2. Let

$$Z_n = \frac{X_1 + X_2 + \cdots + X_n - n\mu}{\sqrt{n}\,\sigma}$$

Then for any interval $\{a \leq x \leq b\}$,

$$\lim_{n \to \infty} P(a \leq Z_n \leq b) = P(a \leq \phi \leq b)$$

where ϕ is the standard normal distribution.

Recall that we called $\bar{S}_n = (X_1 + X_2 + \cdots + X_n)/n$ the sample mean of the random variables $X_1, \ldots, X_n$. Thus Z_n in the above theorem is the standardized sample mean. Roughly speaking, the central limit theorem says that in any sequence of repeated trials the standardized sample mean approaches the standard normal curve as the number of trials increase.

POISSON DISTRIBUTION

The Poisson distribution is defined as follows:

$$p(k; \lambda) = \frac{\lambda^k e^{-\lambda}}{k!}, \quad k = 0, 1, 2, \ldots$$

where $\lambda > 0$ is some constant. This countably infinite distribution appears in many natural phenomena, such as the number of telephone calls per minute at some switchboard, the number of misprints per page in a large text, and the number of α particles emitted by a radioactive substance. Diagrams of the Poisson distribution for various values of λ follow.

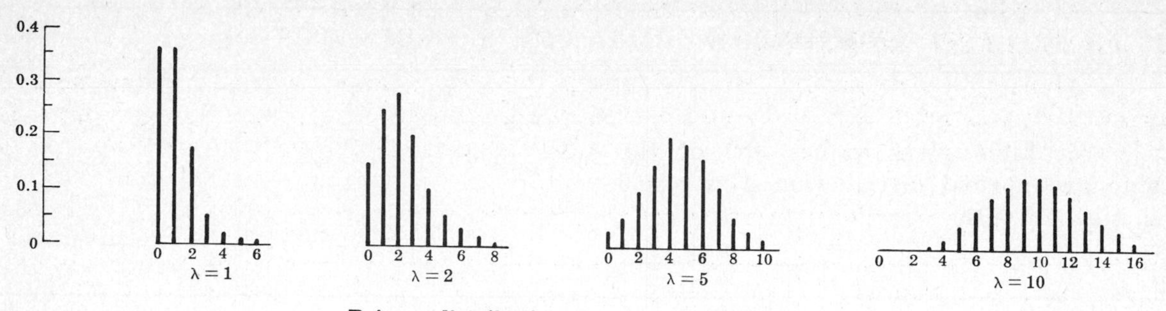

Poisson distribution for selected values of λ

Properties of the Poisson distribution follow:

Theorem 6.5:

Poisson distribution	
Mean	$\mu = \lambda$
Variance	$\sigma^2 = \lambda$
Standard deviation	$\sigma = \sqrt{\lambda}$

Although the Poisson distribution is of independent interest, it also provides us with a close approximation of the binomial distribution for small k provided that p is small and $\lambda = np$ (see Problem 6.27). This is indicated in the following table.

k	0	1	2	3	4	5
Binomial	.366	.370	.185	.0610	.0149	.0029
Poisson	.368	.368	.184	.0613	.0153	.00307

Comparison of Binomial and Poisson distributions with $n = 100$, $p = 1/100$ and $\lambda = np = 1$

MULTINOMIAL DISTRIBUTION

The binomial distribution is generalized as follows. Suppose the sample space of an experiment is partitioned into, say, s mutually exclusive events $A_1, A_2, \ldots, A_s$ with respective probabilities $p_1, p_2, \ldots, p_s$. (Hence $p_1 + p_2 + \cdots + p_s = 1$.) Then:

Theorem 6.6: In n repeated trials, the probability that A_1 occurs k_1 times, A_2 occurs k_2 times, ..., and A_s occurs k_s times is equal to

$$\frac{n!}{k_1!\,k_2!\cdots k_s!}\,p_1^{k_1}\,p_2^{k_2}\cdots p_s^{k_s}$$

where $k_1 + k_2 + \cdots + k_s = n$.

The above numbers form the so-called *multinomial distribution* since they are precisely the terms in the expansion of $(p_1 + p_2 + \cdots + p_s)^n$. Observe that if $s = 2$ then we obtain the binomial distribution, discussed at the beginning of the chapter.

Example 6.4: A fair die is tossed 8 times. The probability of obtaining the faces 5 and 6 twice and each of the others once is

$$\frac{8!}{2!\,2!\,1!\,1!\,1!\,1!}\left(\tfrac{1}{6}\right)^2\left(\tfrac{1}{6}\right)^2\left(\tfrac{1}{6}\right)\left(\tfrac{1}{6}\right)\left(\tfrac{1}{6}\right)\left(\tfrac{1}{6}\right) = \frac{35}{5832} \approx .006$$

STANDARD NORMAL CURVE ORDINATES

This table gives values $\phi(t)$ of the standard normal distribution ϕ at $t \geq 0$ in steps of 0.01.

t	0	1	2	3	4	5	6	7	8	9
0.0	.3989	.3989	.3989	.3988	.3986	.3984	.3982	.3980	.3977	.3973
0.1	.3970	.3965	.3961	.3956	.3951	.3945	.3939	.3932	.3925	.3918
0.2	.3910	.3902	.3894	.3885	.3876	.3867	.3857	.3847	.3836	.3825
0.3	.3814	.3802	.3790	.3778	.3765	.3752	.3739	.3725	.3712	.3697
0.4	.3683	.3668	.3653	.3637	.3621	.3605	.3589	.3572	.3555	.3538
0.5	.3521	.3503	.3485	.3467	.3448	.3429	.3410	.3391	.3372	.3352
0.6	.3332	.3312	.3292	.3271	.3251	.3230	.3209	.3187	.3166	.3144
0.7	.3123	.3101	.3079	.3056	.3034	.3011	.2989	.2966	.2943	.2920
0.8	.2897	.2874	.2850	.2827	.2803	.2780	.2756	.2732	.2709	.2685
0.9	.2661	.2637	.2613	.2589	.2565	.2541	.2516	.2492	.2468	.2444
1.0	.2420	.2396	.2371	.2347	.2323	.2299	.2275	.2251	.2227	.2203
1.1	.2179	.2155	.2131	.2107	.2083	.2059	.2036	.2012	.1989	.1965
1.2	.1942	.1919	.1895	.1872	.1849	.1826	.1804	.1781	.1758	.1736
1.3	.1714	.1691	.1669	.1647	.1626	.1604	.1582	.1561	.1539	.1518
1.4	.1497	.1476	.1456	.1435	.1415	.1394	.1374	.1354	.1334	.1315
1.5	.1295	.1276	.1257	.1238	.1219	.1200	.1182	.1163	.1145	.1127
1.6	.1109	.1092	.1074	.1057	.1040	.1023	.1006	.0989	.0973	.0957
1.7	.0940	.0925	.0909	.0893	.0878	.0863	.0848	.0833	.0818	.0804
1.8	.0790	.0775	.0761	.0748	.0734	.0721	.0707	.0694	.0681	.0669
1.9	.0656	.0644	.0632	.0620	.0608	.0596	.0584	.0573	.0562	.0551
2.0	.0540	.0529	.0519	.0508	.0498	.0488	.0478	.0468	.0459	.0449
2.1	.0440	.0431	.0422	.0413	.0404	.0396	.0387	.0379	.0371	.0363
2.2	.0355	.0347	.0339	.0332	.0325	.0317	.0310	.0303	.0297	.0290
2.3	.0283	.0277	.0270	.0264	.0258	.0252	.0246	.0241	.0235	.0229
2.4	.0224	.0219	.0213	.0208	.0203	.0198	.0194	.0189	.0184	.0180
2.5	.0175	.0171	.0167	.0163	.0158	.0154	.0151	.0147	.0143	.0139
2.6	.0136	.0132	.0129	.0126	.0122	.0119	.0116	.0113	.0110	.0107
2.7	.0104	.0101	.0099	.0096	.0093	.0091	.0088	.0086	.0084	.0081
2.8	.0079	.0077	.0075	.0073	.0071	.0069	.0067	.0065	.0063	.0061
2.9	.0060	.0058	.0056	.0055	.0053	.0051	.0050	.0048	.0047	.0046
3.0	.0044	.0043	.0042	.0040	.0039	.0038	.0037	.0036	.0035	.0034
3.1	.0033	.0032	.0031	.0030	.0029	.0028	.0027	.0026	.0025	.0025
3.2	.0024	.0023	.0022	.0022	.0021	.0020	.0020	.0019	.0018	.0018
3.3	.0017	.0017	.0016	.0016	.0015	.0015	.0014	.0014	.0013	.0013
3.4	.0012	.0012	.0012	.0011	.0011	.0010	.0010	.0010	.0009	.0009
3.5	.0009	.0008	.0008	.0008	.0008	.0007	.0007	.0007	.0007	.0006
3.6	.0006	.0006	.0006	.0005	.0005	.0005	.0005	.0005	.0005	.0004
3.7	.0004	.0004	.0004	.0004	.0004	.0004	.0003	.0003	.0003	.0003
3.8	.0003	.0003	.0003	.0003	.0003	.0002	.0002	.0002	.0002	.0002
3.9	.0002	.0002	.0002	.0002	.0002	.0002	.0002	.0002	.0001	.0001

Table 6.1

STANDARD NORMAL CURVE AREAS

This table gives areas under the standard normal distribution ϕ between 0 and $t \geqq 0$ in steps of 0.01.

t	0	1	2	3	4	5	6	7	8	9
0.0	.0000	.0040	.0080	.0120	.0160	.0199	.0239	.0279	.0319	.0359
0.1	.0398	.0438	.0478	.0517	.0557	.0596	.0636	.0675	.0714	.0754
0.2	.0793	.0832	.0871	.0910	.0948	.0987	.1026	.1064	.1103	.1141
0.3	.1179	.1217	.1255	.1293	.1331	.1368	.1406	.1443	.1480	.1517
0.4	.1554	.1591	.1628	.1664	.1700	.1736	.1772	.1808	.1844	.1879
0.5	.1915	.1950	.1985	.2019	.2054	.2088	.2123	.2157	.2190	.2224
0.6	.2258	.2291	.2324	.2357	.2389	.2422	.2454	.2486	.2518	.2549
0.7	.2580	.2612	.2642	.2673	.2704	.2734	.2764	.2794	.2823	.2852
0.8	.2881	.2910	.2939	.2967	.2996	.3023	.3051	.3078	.3106	.3133
0.9	.3159	.3186	.3212	.3238	.3264	.3289	.3315	.3340	.3365	.3389
1.0	.3413	.3438	.3461	.3485	.3508	.3531	.3554	.3577	.3599	.3621
1.1	.3643	.3665	.3686	.3708	.3729	.3749	.3770	.3790	.3810	.3830
1.2	.3849	.3869	.3888	.3907	.3925	.3944	.3962	.3980	.3997	.4015
1.3	.4032	.4049	.4066	.4082	.4099	.4115	.4131	.4147	.4162	.4177
1.4	.4192	.4207	.4222	.4236	.4251	.4265	.4279	.4292	.4306	.4319
1.5	.4332	.4345	.4357	.4370	.4382	.4394	.4406	.4418	.4429	.4441
1.6	.4452	.4463	.4474	.4484	.4495	.4505	.4515	.4525	.4535	.4545
1.7	.4554	.4564	.4573	.4582	.4591	.4599	.4608	.4616	.4625	.4633
1.8	.4641	.4649	.4656	.4664	.4671	.4678	.4686	.4693	.4699	.4706
1.9	.4713	.4719	.4726	.4732	.4738	.4744	.4750	.4756	.4761	.4767
2.0	.4772	.4778	.4783	.4788	.4793	.4798	.4803	.4808	.4812	.4817
2.1	.4821	.4826	.4830	.4834	.4838	.4842	.4846	.4850	.4854	.4857
2.2	.4861	.4864	.4868	.4871	.4875	.4878	.4881	.4884	.4887	.4890
2.3	.4893	.4896	.4898	.4901	.4904	.4906	.4909	.4911	.4913	.4916
2.4	.4918	.4920	.4922	.4925	.4927	.4929	.4931	.4932	.4934	.4936
2.5	.4938	.4940	.4941	.4943	.4945	.4946	.4948	.4949	.4951	.4952
2.6	.4953	.4955	.4956	.4957	.4959	.4960	.4961	.4962	.4963	.4964
2.7	.4965	.4966	.4967	.4968	.4969	.4970	.4971	.4972	.4973	.4974
2.8	.4974	.4975	.4976	.4977	.4977	.4978	.4979	.4979	.4980	.4981
2.9	.4981	.4982	.4982	.4983	.4984	.4984	.4985	.4985	.4986	.4986
3.0	.4987	.4987	.4987	.4988	.4988	.4989	.4989	.4989	.4990	.4990
3.1	.4990	.4991	.4991	.4991	.4992	.4992	.4992	.4992	.4993	.4993
3.2	.4993	.4993	.4994	.4994	.4994	.4994	.4994	.4995	.4995	.4995
3.3	.4995	.4995	.4995	.4996	.4996	.4996	.4996	.4996	.4996	.4997
3.4	.4997	.4997	.4997	.4997	.4997	.4997	.4997	.4997	.4997	.4998
3.5	.4998	.4998	.4998	.4998	.4998	.4998	.4998	.4998	.4998	.4998
3.6	.4998	.4998	.4999	.4999	.4999	.4999	.4999	.4999	.4999	.4999
3.7	.4999	.4999	.4999	.4999	.4999	.4999	.4999	.4999	.4999	.4999
3.8	.4999	.4999	.4999	.4999	.4999	.4999	.4999	.4999	.4999	.4999
3.9	.5000	.5000	.5000	.5000	.5000	.5000	.5000	.5000	.5000	.5000

Table 6.2

VALUES OF $e^{-\lambda}$										
λ	0.0	0.1	0.2	0.3	0.4	0.5	0.6	0.7	0.8	0.9
$e^{-\lambda}$	1.000	.905	.819	.741	.670	.607	.549	.497	.449	.407
λ	1	2	3	4	5	6	7	8	9	10
$e^{-\lambda}$	.368	.135	.0498	.0183	.00674	.00248	.000912	.000335	.000123	.000045

Table 6.3

Solved Problems

BINOMIAL DISTRIBUTION

6.1. Find (i) $b(2;5,\frac{1}{3})$, (ii) $b(3;6,\frac{1}{2})$, (iii) $b(3;4,\frac{1}{4})$.

Here $b(k;n,p) = \binom{n}{k}p^k q^{n-k}$ where $p+q=1$.

(i) $b(2;5,\frac{1}{3}) = \binom{5}{2}(\frac{1}{3})^2(\frac{2}{3})^3 = \frac{5\cdot 4}{1\cdot 2}(\frac{1}{3})^2(\frac{2}{3})^3 = \frac{80}{243}$.

(ii) $b(3;6,\frac{1}{2}) = \binom{6}{3}(\frac{1}{2})^3(\frac{1}{2})^3 = \frac{6\cdot 5\cdot 4}{1\cdot 2\cdot 3}(\frac{1}{2})^3(\frac{1}{2})^3 = \frac{5}{16}$.

(iii) $b(3;4,\frac{1}{4}) = \binom{4}{3}(\frac{1}{4})^3(\frac{3}{4}) = \binom{4}{1}(\frac{1}{4})^3(\frac{3}{4}) = \frac{4}{1}(\frac{1}{4})^3(\frac{3}{4}) = \frac{3}{64}$.

6.2. A fair coin is tossed three times. Find the probability P that there will appear (i) three heads, (ii) two heads, (iii) one head, (iv) no heads.

Method 1. We obtain the following equiprobable space of eight elements:

$$S = \{\text{HHH}, \text{HHT}, \text{HTH}, \text{HTT}, \text{THH}, \text{THT}, \text{TTH}, \text{TTT}\}$$

(i) Three heads (HHH) occurs only once among the eight sample points; hence $P = \frac{1}{8}$.

(ii) Two heads occurs 3 times (HHT, HTH, and THH); hence $P = \frac{3}{8}$.

(iii) One head occurs 3 times (HTT, THT and TTH); hence $P = \frac{3}{8}$.

(iv) No heads, i.e. three tails (TTT), occurs only once; hence $P = \frac{1}{8}$.

Method 2. Use Theorem 6.1 with $n=3$ and $p=q=\frac{1}{2}$.

(i) Here $k=3$ and $P = b(3;3,\frac{1}{2}) = \binom{3}{3}(\frac{1}{2})^3(\frac{1}{2})^0 = 1\cdot\frac{1}{8}\cdot 1 = \frac{1}{8}$.

(ii) Here $k=2$ and $P = b(2;3,\frac{1}{2}) = \binom{3}{2}(\frac{1}{2})^2(\frac{1}{2}) = 3\cdot\frac{1}{4}\cdot\frac{1}{2} = \frac{3}{8}$.

(iii) Here $k=1$ and $P = b(1;3,\frac{1}{2}) = \binom{3}{1}(\frac{1}{2})^1(\frac{1}{2})^2 = 3\cdot\frac{1}{2}\cdot\frac{1}{4} = \frac{3}{8}$.

(iv) Here $k=0$ and $P = b(0;3,\frac{1}{2}) = \binom{3}{0}(\frac{1}{2})^0(\frac{1}{2})^3 = 1\cdot 1\cdot\frac{1}{8} = \frac{1}{8}$.

6.3. Team A has probability $\frac{2}{3}$ of winning whenever it plays. If A plays 4 games, find the probability that A wins (i) exactly 2 games, (ii) at least 1 game, (iii) more than half of the games.

Here $n=4$, $p=\frac{2}{3}$ and $q=1-p=\frac{1}{3}$.

(i) $P(\text{2 wins}) = b(2;4,\frac{2}{3}) = \binom{4}{2}(\frac{2}{3})^2(\frac{1}{3})^2 = \frac{8}{27}$.

(ii) Here $q^4 = (\frac{1}{3})^4 = \frac{1}{81}$ is the probability that A loses all four games. Then $1 - q^4 = \frac{80}{81}$ is the probability of winning at least one game.

(iii) A wins more than half the games if A wins 3 or 4 games. Hence the required probability is

$$P(3 \text{ wins}) + P(4 \text{ wins}) = \binom{4}{3}(\tfrac{2}{3})^3(\tfrac{1}{3}) + \binom{4}{4}(\tfrac{2}{3})^4 = \tfrac{32}{81} + \tfrac{16}{81} = \tfrac{16}{27}$$

6.4. A family has 6 children. Find the probability P that there are (i) 3 boys and 3 girls, (ii) fewer boys than girls. Assume that the probability of any particular child being a boy is $\frac{1}{2}$.

Here $n = 6$ and $p = q = \frac{1}{2}$.

(i) $P = P(3 \text{ boys}) = \binom{6}{3}(\frac{1}{2})^3(\frac{1}{2})^3 = \frac{20}{64} = \frac{5}{16}$.

(ii) There are fewer boys than girls if there are 0, 1 or 2 boys. Hence

$$P = P(0 \text{ boys}) + P(1 \text{ boy}) + P(2 \text{ boys}) = (\tfrac{1}{2})^6 + \binom{6}{1}(\tfrac{1}{2})(\tfrac{1}{2})^5 + \binom{6}{2}(\tfrac{1}{2})^2(\tfrac{1}{2})^4 = \tfrac{11}{32}$$

6.5. How many dice must be thrown so that there is a better than even chance of obtaining a six?

The probability of not obtaining a six on n dice is $(\frac{5}{6})^n$. Hence we seek the smallest n for which $(\frac{5}{6})^n$ is *less* than $\frac{1}{2}$:

$$(\tfrac{5}{6})^1 = \tfrac{5}{6}; \quad (\tfrac{5}{6})^2 = \tfrac{25}{36}; \quad (\tfrac{5}{6})^3 = \tfrac{125}{216}; \quad \text{but} \quad (\tfrac{5}{6})^4 = \tfrac{625}{1296} < \tfrac{1}{2}$$

Thus 4 dice must be thrown.

6.6. The probability of a man hitting a target is $\frac{1}{4}$. (i) If he fires 7 times, what is the probability P of his hittting the target at least twice? (ii) How many times must he fire so that the probability of his hitting the target at least once is greater than $\frac{2}{3}$?

(i) We seek the sum of the probabilities for $k = 2, 3, 4, 5, 6$ and 7. It is simpler in this case to find the sum of the probabilities for $k = 0$ and 1, i.e. no hits or 1 hit, and then subtract it from 1.

$$P(\text{no hits}) = (\tfrac{3}{4})^7 = \tfrac{2187}{16{,}384}, \qquad P(1 \text{ hit}) = \binom{7}{1}(\tfrac{1}{4})(\tfrac{3}{4})^6 = \tfrac{5103}{16{,}384}$$

Then $P = 1 - \frac{2187}{16{,}384} - \frac{5103}{16{,}384} = \frac{4547}{8192}$.

(ii) The probability of not hitting the target is q^n. Thus we seek the smallest n for which q^n is less than $1 - \frac{2}{3} = \frac{1}{3}$, where $q = 1 - p = 1 - \frac{1}{4} = \frac{3}{4}$. Hence compute successive powers of q until $q^n < \frac{1}{3}$ is obtained:

$$(\tfrac{3}{4})^1 = \tfrac{3}{4} \nless \tfrac{1}{3}; \quad (\tfrac{3}{4})^2 = \tfrac{9}{16} \nless \tfrac{1}{3}; \quad (\tfrac{3}{4})^3 = \tfrac{27}{64} \nless \tfrac{1}{3}; \quad \text{but} \quad (\tfrac{3}{4})^4 = \tfrac{81}{256} < \tfrac{1}{3}$$

In other words, he must fire 4 times.

6.7. Prove Theorem 6.1: The probability of exactly k successes in n repeated trials is $b(k; n, p) = \binom{n}{k} p^k q^{n-k}$.

The sample space of the n repeated trials consists of all ordered n-tuples whose components are either s (success) or f (failure). The event A of k successes consists of all ordered n-tuples of which k components are s and the other $n - k$ components are f. The number of n-tuples in the event A is equal to the number of ways that k letters s can be distributed among the n components of an n-tuple; hence A consists of $\binom{n}{k}$ sample points. Since the probability of each point in A is $p^k q^{n-k}$, we have $P(A) = \binom{n}{k} p^k q^{n-k}$.

6.8. Prove Theorem 6.2: Let X be a random variable with the binomial distribution $b(k; n, p)$. Then (i) $E(X) = np$ and (ii) $\text{Var}(X) = npq$. Hence $\sigma_X = \sqrt{npq}$.

(i) Using $b(k; n, p) = \binom{n}{k} p^k q^{n-k}$, we obtain

$$E(X) = \sum_{k=0}^{n} k \cdot b(k; n, p) = \sum_{k=0}^{n} k \frac{n!}{k!\,(n-k)!} p^k q^{n-k}$$
$$= np \sum_{k=1}^{n} \frac{(n-1)!}{(k-1)!\,(n-k)!} p^{k-1} q^{n-k}$$

(we drop the term $k = 0$ since its value is zero, and we factor out np from each term). We let $s = k-1$ in the above sum. As k runs through the values 1 to n, s runs through the values 0 to $n-1$. Thus

$$E(X) = np \sum_{s=0}^{n-1} \frac{(n-1)!}{s!\,(n-1-s)!} p^s q^{n-1-s} = np \sum_{s=0}^{n-1} b(s; n-1, p) = np$$

since, by the binomial theorem,

$$\sum_{s=0}^{n-1} b(s; n-1, p) = (p+q)^{n-1} = 1^{n-1} = 1$$

(ii) We first compute $E(X^2)$:

$$E(X^2) = \sum_{k=0}^{n} k^2\, b(k; n, p) = \sum_{k=0}^{n} k^2 \frac{n!}{k!\,(n-k)!} p^k q^{n-k}$$
$$= np \sum_{k=1}^{n} k \frac{(n-1)!}{(k-1)!\,(n-k)!} p^{k-1} q^{n-k}$$

Again we let $s = k-1$ and obtain

$$E(X^2) = np \sum_{s=0}^{n-1} (s+1) \frac{(n-1)!}{s!\,(n-1-s)!} p^s q^{n-1-s} = np \sum_{s=0}^{n-1} (s+1)\, b(s; n-1, p)$$

But
$$\sum_{s=0}^{n-1} (s+1)\, b(s; n-1, p) = \sum_{s=0}^{n-1} s \cdot b(s; n-1, p) + \sum_{s=0}^{n-1} b(s; n-1, p)$$
$$= (n-1)p + 1 = np + 1 - p = np + q$$

where we use (i) to obtain $(n-1)p$. Accordingly,

$$E(X^2) = np(np+q) = (np)^2 + npq$$

and
$$\text{Var}(X) = E(X^2) - \mu_X^2 = (np)^2 + npq - (np)^2 = npq$$

Thus the theorem is proved.

6.9. Determine the expected number of boys in a family with 8 children, assuming the sex distribution to be equally probable. What is the probability that the expected number of boys does occur?

The expected number of boys is $E = np = 8 \cdot \frac{1}{2} = 4$. The probability that the family has four boys is

$$b(4; 8, \tfrac{1}{2}) = \binom{8}{4} (\tfrac{1}{2})^4 (\tfrac{1}{2})^4 = \frac{8 \cdot 7 \cdot 6 \cdot 5}{1 \cdot 2 \cdot 3 \cdot 4} (\tfrac{1}{2})^8 = \frac{70}{256} = .27$$

6.10. The probability is 0.02 that an item produced by a factory is defective. A shipment of 10,000 items is sent to its warehouse. Find the expected number E of defective items and the standard deviation σ.

$E = np = (10{,}000)(0.02) = 200.$

$\sigma = \sqrt{npq} = \sqrt{(10{,}000)(0.02)(0.98)} = \sqrt{196} = 14.$

NORMAL DISTRIBUTION

6.11. The mean and standard deviation on an examination are 74 and 12 respectively. Find the scores in standard units of students receiving grades (i) 65, (ii) 74, (iii) 86, (iv) 92.

(i) $t = \dfrac{x-\mu}{\sigma} = \dfrac{65-74}{12} = -0.75$ (iii) $t = \dfrac{x-\mu}{\sigma} = \dfrac{86-74}{12} = 1.0$

(ii) $t = \dfrac{x-\mu}{\sigma} = \dfrac{74-74}{12} = 0$ (iv) $t = \dfrac{x-\mu}{\sigma} = \dfrac{92-74}{12} = 1.5$

6.12. Referring to the preceding problem, find the grades corresponding to standard scores (i) -1, (ii) 0.5, (iii) 1.25, (iv) 1.75.

(i) $x = \sigma t + \mu = (12)(-1) + 74 = 62$ (iii) $x = \sigma t + \mu = (12)(1.25) + 74 = 89$

(ii) $x = \sigma t + \mu = (12)(0.5) + 74 = 80$ (iv) $x = \sigma t + \mu = (12)(1.75) + 74 = 95$

6.13. Let ϕ be the standard normal distribution. Find $\phi(t)$ at (i) $t = 1.63$, (ii) $t = -0.75$, (iii) $t = -2.08$.

(i) In Table 6.1, page 110, look down the first column until the entry 1.6 is reached. Then continue right to column 3. The entry is .1057. Hence $\phi(1.63) = .1057$.

(ii) By symmetry, $\phi(-0.75) = \phi(0.75) = .3011$.

(iii) $\phi(-2.08) = \phi(2.08) = .0459$.

6.14. Let X be a random variable with the standard normal distribution ϕ. Find:

(i) $P(0 \le X \le 1.42)$ (v) $P(-1.79 \le X \le -0.54)$

(ii) $P(-0.73 \le X \le 0)$ (vi) $P(X \ge 1.13)$

(iii) $P(-1.37 \le X \le 2.01)$ (vii) $P(|X| \le 0.5)$

(iv) $P(0.65 \le X \le 1.26)$

(i) $P(0 \le X \le 1.42)$ is equal to the area under the standard normal curve between 0 and 1.42. Thus in Table 6.2, page 111, look down the first column until 1.4 is reached, and then continue right to column 2. The entry is .4222. Hence $P(0 \le X \le 1.42) = .4222$.

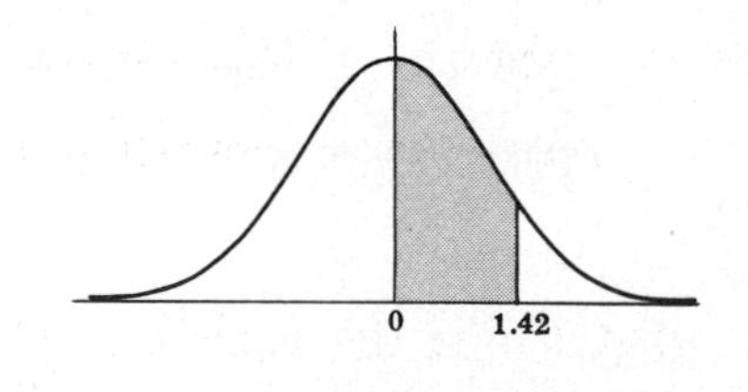

(ii) By symmetry,

$P(-0.73 \le X \le 0)$

$= P(0 \le X \le 0.73) = .2673$

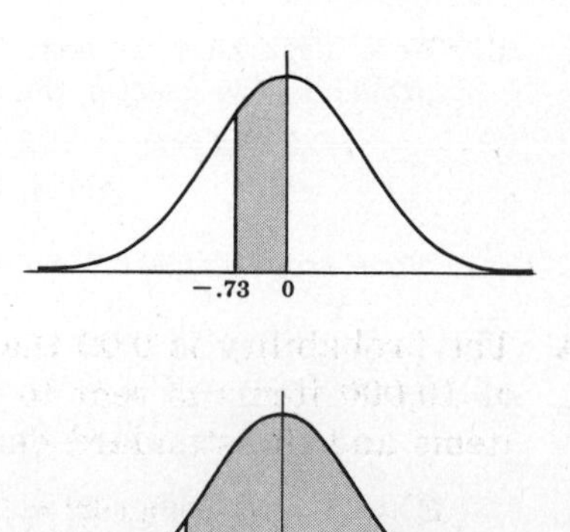

(iii) $P(-1.37 \le X \le 2.01)$

$= P(-1.37 \le X \le 0) + P(0 \le X \le 2.01)$

$= .4147 + .4778 = .8925$

(iv) $P(0.65 \leq X \leq 1.26)$

$= P(0 \leq X \leq 1.26) - P(0 \leq X \leq 0.65)$

$= .3962 - .2422 = .1540$

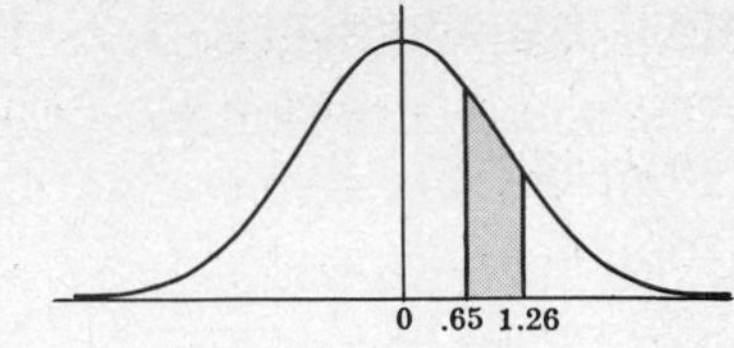

(v) $P(-1.79 \leq X \leq -0.54)$

$= P(0.54 \leq X \leq 1.79)$

$= P(0 \leq X \leq 1.79) - P(0 \leq X \leq 0.54)$

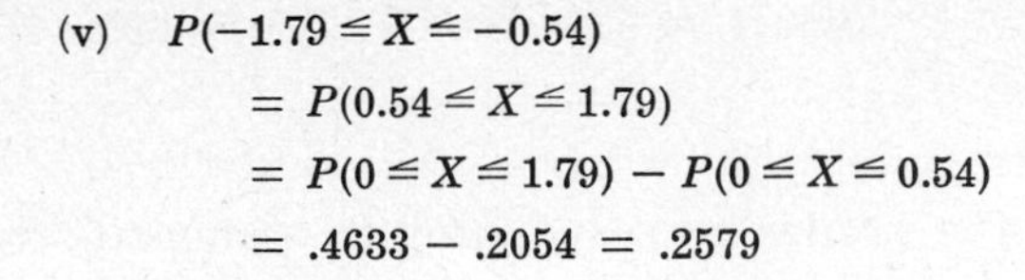

$= .4633 - .2054 = .2579$

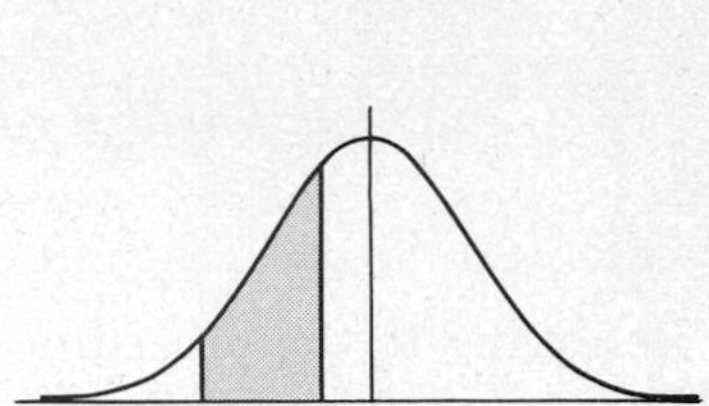

(vi) $P(X \geq 1.13)$

$= P(X \geq 0) - P(0 \leq X \leq 1.13)$

$= .5000 - .3708 = .1292$

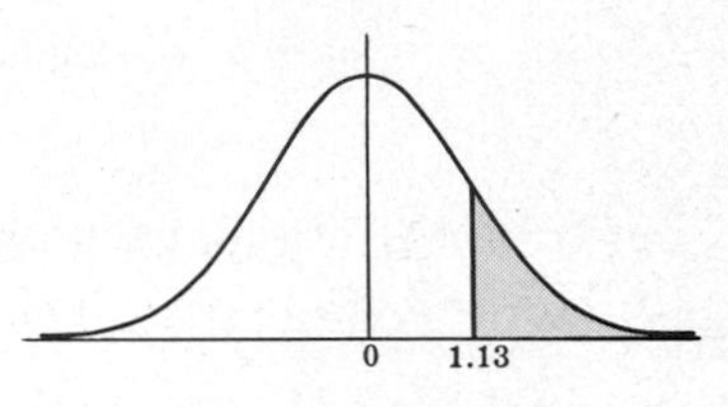

(vii) $P(|X| \leq 0.5)$

$= P(-0.5 \leq X \leq 0.5)$

$= 2P(0 \leq X \leq 0.5)$

$= 2(.1915) = .3830$

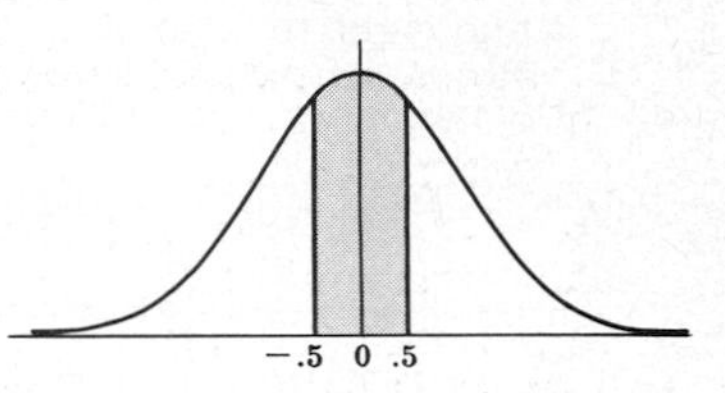

6.15. Let X be a random variable with the standard normal distribution ϕ. Determine the value of t if (i) $P(0 \leq X \leq t) = .4236$, (ii) $P(X \leq t) = .7967$, (iii) $P(t \leq X \leq 2) = .1000$.

(i) In Table 6.2, page 111, the entry .4236 appears to the right of row 1.4 and under column 3. Hence $t = 1.43$.

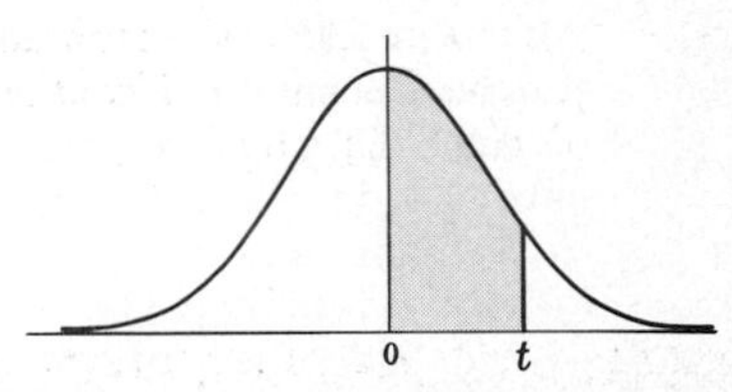

(ii) Note first that t must be positive since the probability is greater than $\frac{1}{2}$. We have

$$P(0 \leq X \leq t) = P(X \leq t) - \tfrac{1}{2}$$
$$= .7967 - .5000 = .2967$$

Thus from Table 6.2 we obtain $t = .83$.

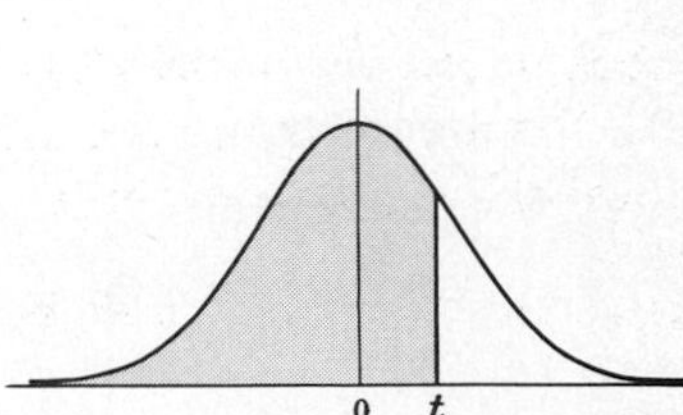

(iii) $P(0 \leq X \leq t) = P(0 \leq X \leq 2) - P(t \leq X \leq 2)$

$$= .4772 - .1000 = .3772$$

Thus from Table 6.2 we obtain $t = 1.161$ (by linear interpolation) or simply $t = 1.16$.

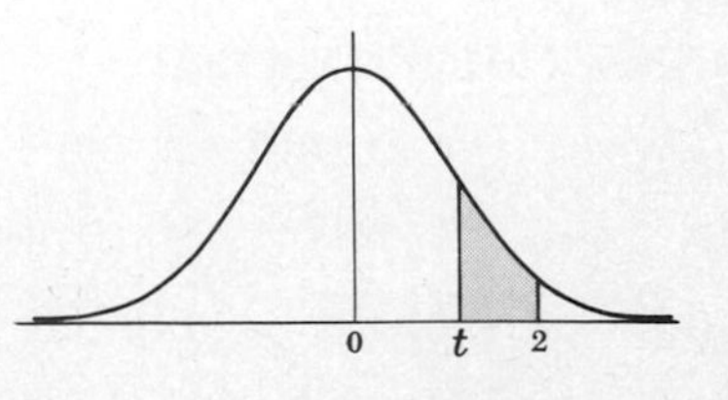

6.16. Suppose the temperature T during June is normally distributed with mean 68° and standard deviation 6°. Find the probability p that the temperature is between 70° and 80°.

$$70° \text{ in standard units } = (70-68)/6 = .33.$$

$$80° \text{ in standard units } = (80-68)/6 = 2.00.$$

Then

$$\begin{aligned} p &= P(70 \leq T \leq 80) = P(.33 \leq T^* \leq 2) \\ &= P(0 \leq T^* \leq 2) - P(0 \leq T^* \leq .33) \\ &= .4772 - .1293 = .3479 \end{aligned}$$

Here T^* is the standardized random variable corresponding to T, and so T^* has the standard normal distribution ϕ.

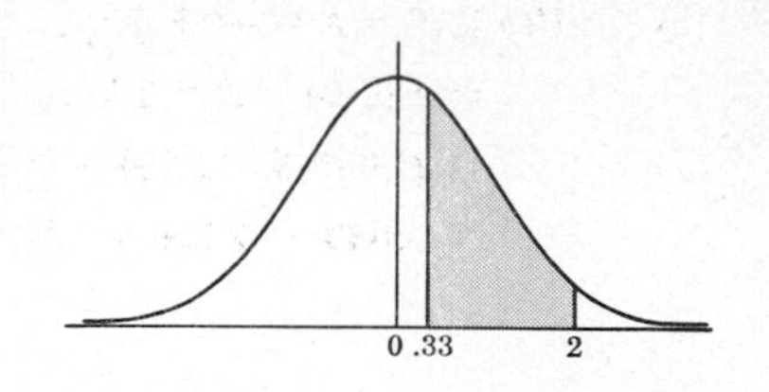

6.17. Suppose the heights H of 800 students are normally distributed with mean 66 inches and standard deviation 5 inches. Find the number N of students with heights (i) between 65 and 70 inches, (ii) greater than or equal to 6 feet (72 inches).

(i) 65 inches in standard units $= (65-66)/5 = -.20$.

70 inches in standard units $= (70-66)/5 = .80$.

Hence

$$\begin{aligned} P(65 \leq H \leq 70) &= P(-.20 \leq H^* \leq .80) \\ &= .0793 + .2881 = .3674 \end{aligned}$$

Thus $N = 800(.3674) = 294$.

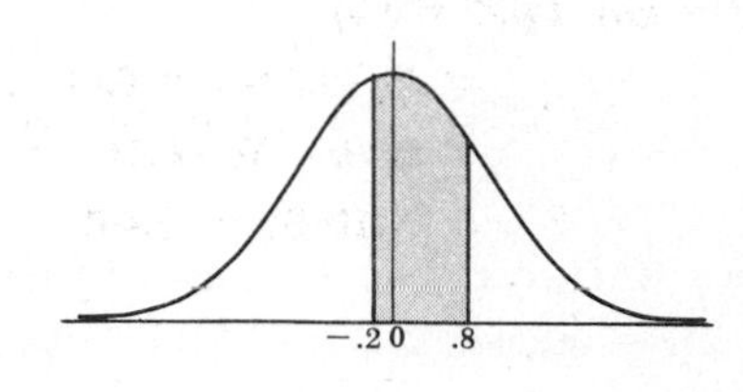

(ii) 72 inches in standard units $= (72-66)/5 = 1.20$.

Hence

$$\begin{aligned} P(H \geq 72) &= P(H^* \geq 1.2) \\ &= .5000 - .3849 = .1151 \end{aligned}$$

Thus $N = 800(.1151) = 92$.

Here H^* is the standardized random variable corresponding to H and so H^* has the standard normal distribution ϕ.

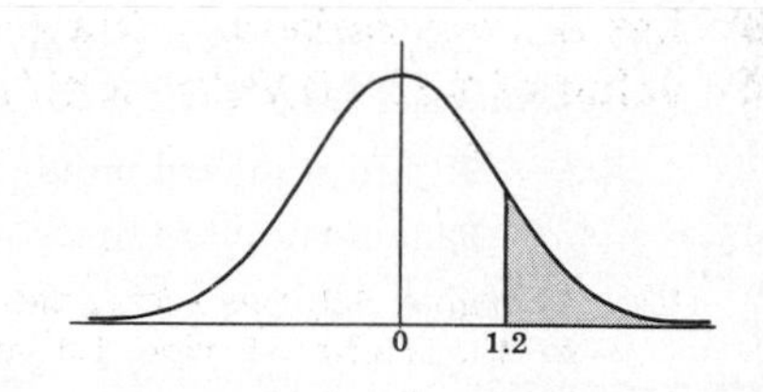

NORMAL APPROXIMATION TO THE BINOMIAL DISTRIBUTION

6.18. A fair coin is tossed 12 times. Determine the probability P that the number of heads occurring is between 4 and 7 inclusive by using (i) the binomial distribution, (ii) the normal approximation to the binomial distribution.

(i) By Theorem 6.1 with $n = 12$ and $p = q = \frac{1}{2}$,

$$P(4 \text{ heads}) = \binom{12}{4}\left(\tfrac{1}{2}\right)^4\left(\tfrac{1}{2}\right)^8 = \tfrac{495}{4096}$$

$$P(5 \text{ heads}) = \binom{12}{5}\left(\tfrac{1}{2}\right)^5\left(\tfrac{1}{2}\right)^7 = \tfrac{792}{4096}$$

$$P(6 \text{ heads}) = \binom{12}{6}\left(\tfrac{1}{2}\right)^6\left(\tfrac{1}{2}\right)^6 = \tfrac{924}{4096}$$

$$P(7 \text{ heads}) = \binom{12}{7}\left(\tfrac{1}{2}\right)^7\left(\tfrac{1}{2}\right)^5 = \tfrac{792}{4096}$$

Hence $P = \frac{495}{4096} + \frac{792}{4096} + \frac{924}{4096} + \frac{792}{4096} = \frac{3003}{4096} = .7332$.

(ii)

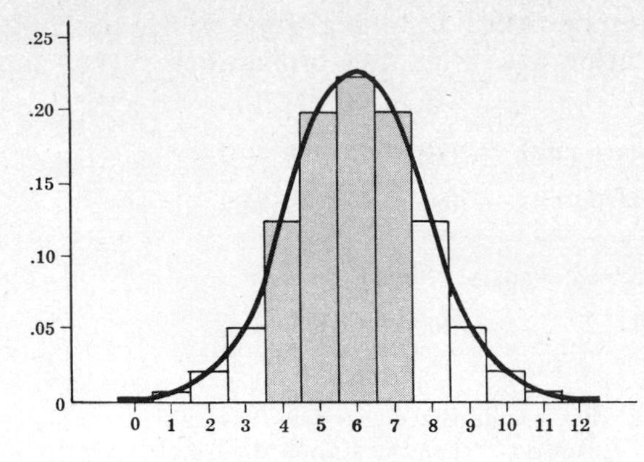

Probability of the number of heads occurring.

Here $\mu = np = 12 \cdot \frac{1}{2} = 6$ and $\sigma = \sqrt{npq} = \sqrt{12 \cdot \frac{1}{2} \cdot \frac{1}{2}} = 1.73$. Let X denote the number of heads occurring. We seek $P(4 \leq X \leq 7)$. But if we assume the data is continuous, in order to apply the normal approximation, we must find $P(3.5 \leq X \leq 7.5)$ as indicated in the above diagram. Now

$$3.5 \text{ in standard units} = (3.5 - 6)/1.73 = -1.45.$$

$$7.5 \text{ in standard units} = (7.5 - 6)/1.73 = .87.$$

Then

$$\begin{aligned} P &\approx P(3.5 \leq X \leq 7.5) \\ &= P(-1.45 \leq X^* \leq .87) \\ &= .4265 + .3078 = .7343 \end{aligned}$$

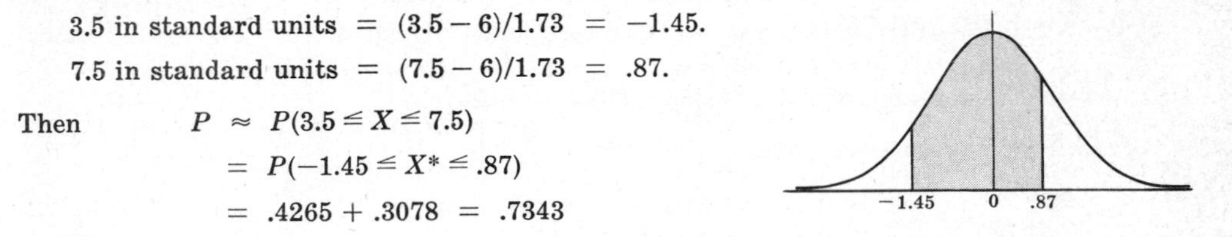

6.19. A fair die is tossed 180 times. Find the probability P that the face 6 will appear (i) between 29 and 32 times inclusive, (ii) between 31 and 35 times inclusive.

Here $\mu = np = 180 \cdot \frac{1}{6} = 30$ and $\sigma = \sqrt{npq} = \sqrt{180 \cdot \frac{1}{6} \cdot \frac{5}{6}} = 5$. Let X denote the number of times the face 6 appears.

(i) We seek $P(29 \leq X \leq 32)$ or, assuming the data is continuous, $P(28.5 \leq X \leq 32.5)$. Now

$$28.5 \text{ in standard units} = (28.5 - 30)/5 = -.3$$

$$32.5 \text{ in standard units} = (32.5 - 30)/5 = .5$$

Hence

$$\begin{aligned} P &\approx P(28.5 \leq X \leq 32.5) = P(-.3 \leq X^* \leq .5) \\ &= P(-.3 \leq X^* \leq 0) + P(0 \leq X^* \leq .5) \\ &= .1179 + .1915 = .3094 \end{aligned}$$

(ii) We seek $P(31 \leq X \leq 35)$ or, assuming the data is continuous, $P(30.5 \leq X \leq 35.5)$. Now

$$30.5 \text{ in standard units} = (30.5 - 30)/5 = .1$$

$$35.5 \text{ in standard units} = (35.5 - 30)/5 = 1.1$$

Then

$$\begin{aligned} P &\approx P(30.5 \leq X \leq 35.5) = P(.1 \leq X^* \leq 1.1) \\ &= P(0 \leq X^* \leq 1.1) - P(0 \leq X^* \leq .1) \\ &= .3643 - .0398 = .3245 \end{aligned}$$

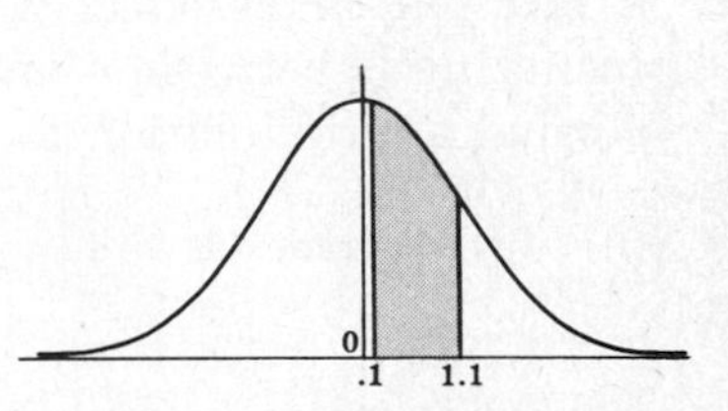

6.20. Among 10,000 random digits, find the probability P that the digit 3 appears at most 950 times.

Here $\mu = np = 10{,}000 \cdot \frac{1}{10} = 1000$ and $\sigma = \sqrt{npq} = \sqrt{10{,}000 \cdot \frac{1}{10} \cdot \frac{9}{10}} = 30$. Let X denote the number of times the digit 3 appears. We seek $P(X \leq 950)$. Now

$$950 \text{ in standard units } = (950 - 1000)/30 = -1.67$$

Thus

$$\begin{aligned} P &\approx P(X \leq 950) = P(X^* \leq -1.67) \\ &= P(X^* \leq 0) - P(-1.67 \leq X^* \leq 0) \\ &= .5000 - .4525 = .0475 \end{aligned}$$

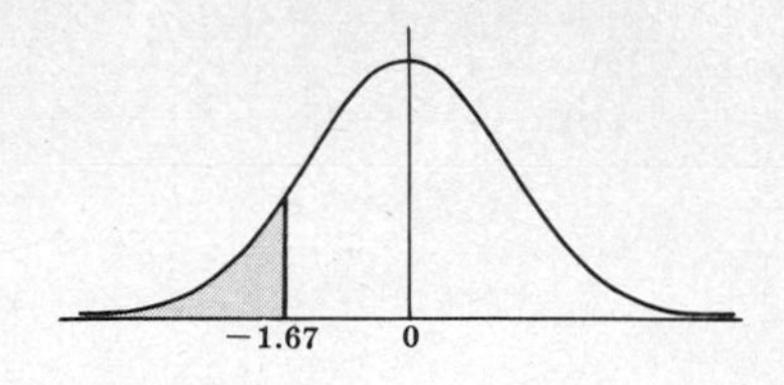

POISSON DISTRIBUTION

6.21. Find: (i) $e^{-1.3}$, (ii) $e^{-2.5}$.

By Table 6.3, page 112, and the law of exponents:

(i) $e^{-1.3} = (e^{-1})(e^{-0.3}) = (.368)(.741) = .273.$

(ii) $e^{-2.5} = (e^{-2})(e^{-0.5}) = (.135)(.607) = .0819.$

6.22. For the Poisson distribution $p(k;\lambda) = \dfrac{\lambda^k e^{-\lambda}}{k!}$, find (i) $p(2;1)$, (ii) $p(3;\frac{1}{2})$, (iii) $p(2;.7)$.

(Use Table 6.3, page 112, to obtain $e^{-\lambda}$.)

(i) $p(2;1) = \dfrac{1^2 e^{-1}}{2!} = \dfrac{e^{-1}}{2} = \dfrac{.368}{2} = .184.$

(ii) $p(3;\frac{1}{2}) = \dfrac{(\frac{1}{2})^3 e^{-.5}}{3!} = \dfrac{e^{-.5}}{48} = \dfrac{.607}{48} = .013.$

(iii) $p(2;.7) = \dfrac{(.7)^2 e^{-.7}}{2!} = \dfrac{(.49)(.497)}{2} = .12.$

6.23. Suppose 300 misprints are distributed randomly throughout a book of 500 pages. Find the probability P that a given page contains (i) exactly 2 misprints, (ii) 2 or more misprints.

We view the number of misprints on one page as the number of successes in a sequence of Bernoulli trials. Here $n = 300$ since there are 300 misprints, and $p = 1/500$, the probability that a misprint appears on the given page. Since p is small, we use the Poisson approximation to the binomial distribution with $\lambda = np = 0.6$.

(i) $P = p(2; 0.6) = \dfrac{(.6)^2 e^{-0.6}}{0!} = (.36)(.549)/2 = .0988 \approx .1.$

(ii) $P(0 \text{ misprints}) = \dfrac{(.6)^0 e^{-0.6}}{0!} = e^{-0.6} = .549$

$P(1 \text{ misprint}) = \dfrac{(.6)e^{-0.6}}{1!} = (.6)(.549) = .329$

Then $P = 1 - P(0 \text{ or } 1 \text{ misprint}) = 1 - (.549 + .329) = .122.$

6.24. Suppose 2% of the items made by a factory are defective. Find the probability P that there are 3 defective items in a sample of 100 items.

The binomial distribution with $n = 100$ and $p = .02$ applies. However, since p is small, we use the Poisson approximation with $\lambda = np = 2$. Thus

$$P = p(3;2) = \frac{2^3 e^{-2}}{3!} = 8(.135)/6 = .180$$

6.25. Show that the Poisson distribution $p(k; \lambda)$ is a probability distribution, i.e.

$$\sum_{k=0}^{\infty} p(k; \lambda) = 1$$

By known results of analysis, $e^{\lambda} = \sum_{k=0}^{\infty} \lambda^k/k!$. Hence

$$\sum_{k=0}^{\infty} p(k; \lambda) = \sum_{k=0}^{\infty} \frac{\lambda^k e^{-\lambda}}{k!} = e^{-\lambda} \sum_{k=0}^{\infty} \lambda^k/k! = e^{-\lambda} e^{\lambda} = 1$$

6.26. Prove Theorem 6.5: Let X be a random variable with the Poisson distribution $p(k; \lambda)$. Then (i) $E(X) = \lambda$ and (ii) $\text{Var}(X) = \lambda$. Hence $\sigma_X = \sqrt{\lambda}$.

(i) Using $p(k; \lambda) = \lambda^k e^{-\lambda}/k!$, we obtain

$$E(X) = \sum_{k=0}^{\infty} k \cdot p(k; \lambda) = \sum_{k=0}^{\infty} k \frac{\lambda^k e^{-\lambda}}{k!} = \lambda \sum_{k=1}^{\infty} \frac{\lambda^{k-1} e^{-\lambda}}{(k-1)!}$$

(we drop the term $k = 0$ since its value is zero, and we factor out λ from each term). Let $s = k - 1$ in the above sum. As k runs through the values 1 to ∞, s runs through the values 0 to ∞. Thus

$$E(X) = \lambda \sum_{s=0}^{\infty} \frac{\lambda^s e^{-\lambda}}{s!} = \lambda \sum_{s=0}^{\infty} p(s; \lambda) = \lambda$$

since $\sum_{s=0}^{\infty} p(s; \lambda) = 1$ by the preceding problem.

(ii) We first compute $E(X^2)$:

$$E(X^2) = \sum_{k=0}^{\infty} k^2 p(k; \lambda) = \sum_{k=0}^{\infty} k^2 \frac{\lambda^k e^{-\lambda}}{k!} = \lambda \sum_{k=1}^{\infty} k \frac{\lambda^{k-1} e^{-\lambda}}{(k-1)!}$$

Again we let $s = k - 1$ and obtain

$$E(X^2) = \lambda \sum_{s=0}^{\infty} (s+1) \frac{\lambda^s e^{-\lambda}}{s!} = \lambda \sum_{s=0}^{\infty} (s+1)\, p(s; \lambda)$$

But

$$\sum_{s=0}^{\infty} (s+1)\, p(s; \lambda) = \sum_{s=0}^{\infty} s p(s; \lambda) + \sum_{s=0}^{\infty} p(s; \lambda) = \lambda + 1$$

where we use (i) to obtain λ and the preceding problem to obtain 1. Accordingly,

$$E(X^2) = \lambda(\lambda + 1) = \lambda^2 + \lambda$$

and

$$\text{Var}(X) = E(X^2) - \mu_X^2 = \lambda^2 + \lambda - \lambda^2 = \lambda$$

Thus the theorem is proved.

6.27. Show that if p is small and n is large, then the binomial distribution is approximated by the Poisson distribution; that is, $b(k; n, p) \approx p(k; \lambda)$ where $\lambda = np$.

We have $b(0; n, p) = (1-p)^n = (1 - \lambda/n)^n$. Taking the natural logarithm of both sides,

$$\ln b(0; n, p) = n \ln(1 - \lambda/n)$$

The Taylor expansion of the natural logarithm is

$$\ln(1 + x) = x - \frac{x^2}{2} + \frac{x^3}{3} - \cdots$$

and so

$$\ln\left(1 - \frac{\lambda}{n}\right) = -\frac{\lambda}{n} - \frac{\lambda^2}{2n^2} - \frac{\lambda^3}{3n^3} - \cdots$$

Therefore if n is large,

$$\ln b(0; n, p) \;=\; n \ln\left(1 - \frac{\lambda}{n}\right) \;=\; -\lambda - \frac{\lambda^2}{2n} - \frac{\lambda^3}{3n^2} \;\approx\; -\lambda$$

and hence $b(0; n, p) \approx e^{-\lambda}$.

Furthermore if p is very small and hence $q \approx 1$, we have

$$\frac{b(k; n, p)}{b(k-1; n, p)} \;=\; \frac{(n-k+1)p}{kq} \;=\; \frac{\lambda - (k-1)p}{kq} \;\approx\; \frac{\lambda}{k}$$

That is, $b(k; n, p) \approx \frac{\lambda}{k} b(k-1; n, p)$. Thus using $b(0; n, p) \approx e^{-\lambda}$, we obtain $b(1; n, p) \approx \lambda e^{-\lambda}$, $b(2; n, p) \approx \lambda^2 e^{-\lambda}/2$ and, by induction, $b(k; n, p) \approx \frac{\lambda^k e^{-\lambda}}{k!} = p(k; \lambda)$.

MISCELLANEOUS PROBLEMS

6.28. The painted light bulbs produced by a company are 50% red, 30% blue and 20% green. In a sample of 5 bulbs, find the probability P that 2 are red, 1 is green and 2 are blue.

By Theorem 6.6 on the multinomial distribution,

$$P \;=\; \frac{5!}{2!\,1!\,2!} (.5)^2 (.3) (.2)^2 \;=\; .09$$

6.29. Show that the normal distribution

$$f(x) \;=\; \frac{1}{\sigma\sqrt{2\pi}}\, e^{\frac{1}{2}(x-\mu)^2/\sigma^2}$$

is a continuous probability distribution, i.e. $\int_{-\infty}^{\infty} f(x)\, dx \;=\; 1.$

Substituting $t = (x-\mu)/\sigma$ in $\int_{-\infty}^{\infty} f(x)\, dx$, we obtain the integral

$$I \;=\; \frac{1}{\sqrt{2\pi}} \int_{-\infty}^{\infty} e^{-t^2/2}\, dt$$

It suffices to show that $I^2 = 1$. We have

$$I^2 \;=\; \frac{1}{2\pi} \int_{-\infty}^{\infty} e^{-t^2/2}\, dt \int_{-\infty}^{\infty} e^{-s^2/2}\, ds \;=\; \frac{1}{2\pi} \int_{-\infty}^{\infty} \int_{-\infty}^{\infty} e^{-(s^2+t^2)/2}\, ds\, dt$$

We introduce polar coordinates in the above double integral. Let $s = r\cos\theta$ and $t = r\sin\theta$. Then $ds\, dt = r\, dr\, d\theta$, and $0 \leq \theta \leq 2\pi$ and $0 \leq r \leq \infty$. That is,

$$I^2 \;=\; \frac{1}{2\pi} \int_0^{2\pi} \int_0^{\infty} r e^{-r^2/2}\, dr\, d\theta$$

But

$$\int_0^{\infty} r e^{-r^2/2}\, dr \;=\; \left[-e^{-r^2/2}\right]_0^{\infty} \;=\; 1$$

Hence $I^2 = \frac{1}{2\pi} \int_0^{2\pi} d\theta = 1$ and the theorem is proved.

6.30. Prove Theorem 6.3: Let X be a random variable with the normal distribution

$$f(x) \;=\; \frac{1}{\sigma\sqrt{2\pi}}\, e^{-\frac{1}{2}(x-\mu)^2/\sigma^2}$$

Then (i) $E(X) = \mu$ and (ii) $\mathrm{Var}(X) = \sigma^2$. Hence $\sigma_X = \sigma$.

(i) By definition, $E(X) = \frac{1}{\sigma\sqrt{2\pi}} \int_{-\infty}^{\infty} x e^{-\frac{1}{2}(x-\mu)^2/\sigma^2}\, dx$. Setting $t = (x-\mu)/\sigma$, we obtain

$$E(X) = \frac{1}{\sqrt{2\pi}}\int_{-\infty}^{\infty}(\sigma t + \mu)e^{-t^2/2}\,dt = \frac{\sigma}{\sqrt{2\pi}}\int_{-\infty}^{\infty} te^{-t^2/2}\,dt + \mu\frac{1}{\sqrt{2\pi}}\int_{-\infty}^{\infty} e^{-t^2/2}\,dt$$

But $g(t) = te^{-t^2/2}$ is an odd function, i.e. $g(-t) = -g(t)$; hence $\int_{-\infty}^{\infty} te^{-t^2/2}\,dt = 0$. Furthermore, $\frac{1}{\sqrt{2\pi}}\int_{-\infty}^{\infty} e^{-t^2/2}\,dt = 1$ by the preceding problem. Accordingly, $E(X) = \frac{\sigma}{\sqrt{2\pi}}\cdot 0 + \mu\cdot 1 = \mu$ as claimed.

(ii) By definition, $E(X^2) = \frac{1}{\sigma\sqrt{2\pi}}\int_{-\infty}^{\infty} x^2 e^{-\frac{1}{2}(x-\mu)^2/\sigma^2}\,dx$. Again setting $t = (x-\mu)/\sigma$, we obtain

$$E(X^2) = \frac{1}{\sqrt{2\pi}}\int_{-\infty}^{\infty}(\sigma t + \mu)^2 e^{-t^2/2}\,dt$$

$$= \sigma^2\frac{1}{\sqrt{2\pi}}\int_{-\infty}^{\infty} t^2e^{-t^2/2}\,dt + 2\mu\sigma\frac{1}{\sqrt{2\pi}}\int_{-\infty}^{\infty} te^{-t^2/2}\,dt + \mu^2\frac{1}{\sqrt{2\pi}}\int_{-\infty}^{\infty} e^{-t^2/2}\,dt$$

which reduces as above to $E(X^2) = \sigma^2\frac{1}{\sqrt{2\pi}}\int_{-\infty}^{\infty} t^2e^{-t^2/2}\,dt + \mu^2$

We integrate the above integral by parts. Let $u = t$ and $dv = te^{-t^2/2}\,dt$. Then $v = -e^{-t^2/2}$ and $du = dt$. Thus

$$\frac{1}{\sqrt{2\pi}}\int_{-\infty}^{\infty} t^2e^{-t^2/2}\,dt = \frac{1}{\sqrt{2\pi}}\Big[-te^{-t^2/2}\Big]_{-\infty}^{\infty} + \frac{1}{\sqrt{2\pi}}\int_{-\infty}^{\infty} e^{-t^2/2}\,dt = 0 + 1 = 1$$

Consequently, $E(X^2) = \sigma^2\cdot 1 + \mu^2 = \sigma^2 + \mu^2$ and

$$\mathrm{Var}(X) = E(X^2) - \mu_X^2 = \sigma^2 + \mu^2 - \mu^2 = \sigma^2$$

Thus the theorem is proved.

Supplementary Problems

BINOMIAL DISTRIBUTION

6.31. Find (i) $b(1; 5, \frac{1}{3})$, (ii) $b(2; 7, \frac{1}{2})$, (iii) $b(2; 4, \frac{1}{4})$.

6.32. A card is drawn and replaced three times from an ordinary deck of 52 cards. Find the probability that (i) two hearts are drawn, (ii) three hearts are drawn, (iii) at least one heart is drawn.

6.33. A baseball player's batting average is .300. If he comes to bat 4 times, what is the probability that he will get (i) two hits, (ii) at least one hit?

6.34. A box contains 3 red marbles and 2 white marbles. A marble is drawn and replaced three times from the box. Find the probability that (i) 1 red marble was drawn, (ii) 2 red marbles were drawn, (iii) at least one red marble was drawn.

6.35. Team A has probability $\frac{2}{5}$ of winning whenever it plays. If A plays 4 games, find the probability that A wins (i) 2 games, (ii) at least 1 game, (iii) more than half of the games.

6.36. A card is drawn and replaced in an ordinary deck of 52 cards. How many times must a card be drawn so that (i) there is at least an even chance of drawing a heart, (ii) the probability of drawing a heart is greater than $\frac{3}{4}$?

6.37. The probability of a man hitting a target is $\frac{1}{3}$. (i) If he fires 5 times, what is the probability of hitting the target at least twice? (ii) How many times must he fire so that the probability of hitting the target at least once is more than 90%.

6.38. The mathematics department has 8 graduate assistants who are assigned to the same office. Each assistant is just as likely to study at home as in the office. How many desks must there be in the office so that each assistant has a desk at least 90% of the time?

6.39. Of the bolts produced by a factory, 2% are defective. In a shipment of 3600 bolts from the factory, find the expected number of defective bolts and the standard deviation.

6.40. A fair die is tossed 1620 times. Find the expected number of times the face 6 occurs and the standard deviation.

6.41. Let X be a binomially distributed random variable with $E(X) = 2$ and $\text{Var}(X) = 4/3$. Find the distribution of X.

6.42. Consider the binomial distribution $P(k) = b(k; n, p)$. Show that:

(i) $\dfrac{P(k)}{P(k-1)} = \dfrac{b(k; n, p)}{b(k-1; n, p)} = \dfrac{(n-k+1)p}{kq}$

(ii) $P(k) > P(k-1)$ for $k < (n+1)p$ and $P(k) < P(k-1)$ for $k > (n+1)p$.

NORMAL DISTRIBUTION

6.43. Let ϕ be the standard normal distribution.

(i) Find $\phi(\frac{1}{4})$, $\phi(\frac{1}{2})$ and $\phi(-\frac{3}{4})$.

(ii) Find t such that (*a*) $\phi(t) = .100$, (*b*) $\phi(t) = .2500$, (*c*) $\phi(t) = .4500$.

6.44. Let X be a random variable with the standard normal distribution ϕ. Find:
(i) $P(-.81 \leq X \leq 1.13)$, (ii) $P(.53 \leq X \leq 2.03)$, (iii) $P(X \leq .73)$, (iv) $P(|X| \leq \frac{1}{4})$.

6.45. Let X be normally distributed with mean 8 and standard deviation 4. Find:
(i) $P(5 \leq X \leq 10)$, (ii) $P(10 \leq X \leq 15)$, (iii) $P(X \geq 15)$, (iv) $P(X \leq 5)$.

6.46. Suppose the weights of 2000 male students are normally distributed with mean 155 pounds and standard deviation 20 pounds. Find the number of students with weights (i) less than or equal to 100 pounds, (ii) between 120 and 130 pounds, (iii) between 150 and 175 pounds, (iv) greater than or equal to 200 pounds.

6.47. Suppose the diameters of bolts manufactured by a company are normally distributed with mean .25 inches and standard deviation .02 inches. A bolt is considered defective if its diameter is $\leq .20$ inches or $\geq .28$ inches. Find the percentage of defective bolts manufactured by the company.

6.48. Suppose the scores on an examination are normally distributed with mean 76 and standard deviation 15. The top 15% of the students receive A's and the bottom 10% receive F's. Find (i) the minimum score to receive an A and (ii) the minimum score to pass (not to receive an F).

NORMAL APPROXIMATION TO THE BINOMIAL DISTRIBUTION

6.49. A fair coin is tossed 10 times. Find the probability of obtaining between 4 and 7 heads inclusive by using (i) the binomial distribution, (ii) the normal approximation to the binomial distribution.

6.50. A fair coin is tossed 400 times. Find the probability that the number of heads which occur differs from 200 by (i) more than 10, (ii) more than 25.

6.51. A fair die is tossed 720 times. Find the probability that the face 6 will occur (i) between 100 and 125 times inclusive, (ii) more than 150 times.

6.52. Among 625 random digits, find the probability that the digit 7 appears (i) between 50 and 60 times inclusive, (ii) between 60 and 70 times inclusive.

POISSON DISTRIBUTION

6.53. Find (i) $e^{-1.6}$, (ii) $e^{-2.3}$.

6.54. For the Poisson distribution $p(k; \lambda)$, find (i) $p(2; 1.5)$, (ii) $p(3; 1)$, (iii) $p(2; .6)$.

6.55. Suppose 220 misprints are distributed randomly throughout a book of 200 pages. Find the probability that a given page contains (i) no misprints, (ii) 1 misprint, (iii) 2 misprints, (iv) 2 or more misprints.

6.56. Suppose 1% of the items made by a machine are defective. Find the probability that 3 or more items are defective in a sample of 100 items.

6.57. Suppose 2% of the people on the average are left-handed. Find the probability of 3 or more left-handed among 100 people.

6.58. Suppose there is an average of 2 suicides per year per 50,000 population. In a city of 100,000 find the probability that in a given year there are (i) 0, (ii) 1, (iii) 2, (iv) 2 or more suicides.

MULTINOMIAL DISTRIBUTION

6.59. A die is "loaded" so that the face 6 appears .3 of the time, the opposite face 1 appears .1 of the time, and each of the other faces appears .15 of the time. The die is tossed 6 times. Find the probability that (i) each face appears once, (ii) the faces 4, 5 and 6 each appears twice.

6.60. A box contains 5 red, 3 white and 2 blue marbles. A sample of six marbles is drawn with replacement, i.e. each marble is replaced before the next one is drawn. Find the probability that: (i) 3 are red, 2 are white and 1 is blue; (ii) 2 are red, 3 are white and 1 is blue; (iii) 2 of each color appears.

Answers to Supplementary Problems

6.31. (i) $\frac{80}{243}$, (ii) $\frac{21}{128}$, (iii) $\frac{27}{128}$

6.32. (i) $\frac{9}{64}$, (ii) $\frac{1}{64}$, (iii) $\frac{37}{64}$

6.33. (i) 0.2646, (ii) 0.7599

6.34. (i) $\frac{36}{125}$, (ii) $\frac{54}{125}$, (iii) $\frac{117}{125}$

6.35. (i) $\frac{216}{625}$, (ii) $\frac{544}{625}$, (iii) $\frac{112}{625}$

6.36. (i) 3, (ii) 5

6.37. (i) $\frac{131}{243}$, (ii) 6

6.38. 6

6.39. $\mu = 72$, $\sigma = 8.4$

6.40. $\mu = 270$, $\sigma = 15$

6.41.

x_i	0	1	2	3	4	5	6
$f(x_i)$	64/729	192/729	240/729	160/729	60/729	12/729	1/729

Distribution of X with $n = 6$ and $p = 1/3$

6.43. (i) $\phi(\frac{1}{4}) = .3867$, $\phi(\frac{1}{2}) = .3521$, $\phi(-\frac{3}{4}) = .3011$.
(ii) (*a*) $t = \pm 1.66$, (*b*) $t = \pm .97$, (*c*) there is no value of t.

6.44. (i) $.2910 + .3708 = .6618$, (ii) $.4788 - .2019 = .2769$, (iii) $.5000 + .2673 = .7673$, (iv) $2(.0987) = .1974$.

6.45. (i) .4649, (ii) .2684, (iii) .0401, (iv) .2266

6.46. (i) 6, (ii) 131, (iii) 880, (iv) 24

6.47. 7.3%

6.48. (i) 92, (ii) 57

6.49. (i) .7734, (ii) .7718

6.50. (i) .2938, (ii) .0108

6.51. (i) .6886, (ii) .0011

6.52. (i) .3518, (ii) .5131

6.53. (i) .202, (ii) .100

6.54. (i) .251, (ii) .0613, (iii) .0988

6.55. (i) .333, (ii) .366, (iii) .201, (iv) .301

6.56. 0.080

6.57. 0.325

6.58. (i) .0183, (ii) .0732, (iii) .1464, (iv) .909

6.59. (i) .0109, (ii) .00103

6.60. (i) .135, (ii) .0810, (iii) .0810

Chapter 7

Markov Chains

INTRODUCTION

We review the definitions and elementary properties of vectors and matrices which are required for this chapter.

By a *vector* u we simply mean an n-tuple of numbers:

$$u = (u_1, u_2, \ldots, u_n)$$

The u_i are called the *components* of u. If all the $u_i = 0$, then u is called the *zero vector*. By a *scalar multiple* ku of u (where k is a real number), we mean the vector obtained from u by multiplying its components by k:

$$ku = (ku_1, ku_2, \ldots, ku_n)$$

We note that two vectors are equal if and only if their corresponding components are equal.

By a *matrix* A we mean a rectangular array of numbers:

$$A = \begin{pmatrix} a_{11} & a_{12} & \cdots & a_{1n} \\ a_{21} & a_{22} & \cdots & a_{2n} \\ \cdots & \cdots & \cdots & \cdots \\ a_{m1} & a_{m2} & \cdots & a_{mn} \end{pmatrix}$$

The m horizontal n-tuples

$$(a_{11}, a_{12}, \ldots, a_{1n}),\ (a_{21}, a_{22}, \ldots, a_{2n}),\ \ldots,\ (a_{m1}, a_{m2}, \ldots, a_{mn})$$

are called the *rows* of A, and the n vertical m-tuples

$$\begin{pmatrix} a_{11} \\ a_{21} \\ \cdots \\ a_{m1} \end{pmatrix}, \begin{pmatrix} a_{12} \\ a_{22} \\ \cdots \\ a_{m2} \end{pmatrix}, \ldots, \begin{pmatrix} a_{1n} \\ a_{2n} \\ \cdots \\ a_{mn} \end{pmatrix}$$

its *columns*. Note that the element a_{ij}, called the *ij-entry*, appears in the ith row and the jth column. We frequently denote such a matrix simply by $A = (a_{ij})$.

A matrix with m rows and n columns is said to be an m by n matrix, written $m \times n$ matrix; if $m = n$, then it is called a *square* matrix (or: *n-square* matrix). We also note that a matrix with only one row may be viewed as a vector, and vice versa.

Now suppose A and B are two matrices such that the number of columns of A is equal to the number of rows of B, say A is an $m \times p$ matrix and B is a $p \times n$ matrix. Then the product of A and B, written AB, is the $m \times n$ matrix whose ij-entry is obtained by multiplying the elements of the ith row of A by the corresponding elements of the jth column of B and then adding:

$$\begin{pmatrix} a_{11} & \cdots & a_{1p} \\ \cdot & \cdots & \cdot \\ a_{i1} & \cdots & a_{ip} \\ \cdot & \cdots & \cdot \\ a_{m1} & \cdots & a_{mp} \end{pmatrix} \begin{pmatrix} b_{11} & \cdots & b_{1j} & \cdots & b_{1n} \\ \cdot & \cdots & \cdot & \cdots & \cdot \\ \cdot & \cdots & \cdot & \cdots & \cdot \\ \cdot & \cdots & \cdot & \cdots & \cdot \\ b_{p1} & \cdots & b_{pj} & \cdots & b_{pn} \end{pmatrix} = \begin{pmatrix} c_{11} & \cdots & c_{1n} \\ \cdot & \cdots & \cdot \\ \cdot & c_{ij} & \cdot \\ \cdot & \cdots & \cdot \\ c_{m1} & \cdots & c_{mn} \end{pmatrix}$$

where
$$c_{ij} = a_{i1}b_{1j} + a_{i2}b_{2j} + \cdots + a_{ip}b_{pj} = \sum_{k=1}^{p} a_{ik}b_{kj}$$

If the number of columns of A is not equal to the number of rows of B, say A is $m \times p$ and B is $q \times n$ where $p \neq q$, then the product AB is not defined.

There are special cases of matrix multiplication which are of special interest. If A is an n-square matrix, then we can form all the *powers* of A:

$$A^2 = AA, \quad A^3 = AA^2, \quad A^4 = AA^3, \quad \ldots$$

In addition, if u is a vector with n components, then we can form the product

$$uA$$

which is again a vector with n components. We call $u \neq 0$ a *fixed vector* (or: *fixed point*) of A, if u is "left fixed", i.e. is not changed, when multiplied by A:

$$uA = u$$

In this case, for any scalar $k \neq 0$, we have

$$(ku)A = k(uA) = ku$$

That is,

Theorem 7.1: If u is a fixed vector of a matrix A, then every nonzero scalar multiple ku of u is also a fixed vector of A.

Example 7.1:
$$\begin{pmatrix} r & s \\ t & u \end{pmatrix}\begin{pmatrix} a_1 & a_2 & a_3 \\ b_1 & b_2 & b_3 \end{pmatrix} = \begin{pmatrix} ra_1 + sb_1 & ra_2 + sb_2 & ra_3 + sb_3 \\ ta_1 + ub_1 & ta_2 + ub_2 & ta_3 + ub_3 \end{pmatrix}$$

Example 7.2: If $A = \begin{pmatrix} 1 & 2 \\ 3 & 4 \end{pmatrix}$, then
$$A^2 = \begin{pmatrix} 1 & 2 \\ 3 & 4 \end{pmatrix}\begin{pmatrix} 1 & 2 \\ 3 & 4 \end{pmatrix} = \begin{pmatrix} 1+6 & 2+8 \\ 3+12 & 6+16 \end{pmatrix} = \begin{pmatrix} 7 & 10 \\ 15 & 22 \end{pmatrix}$$

Example 7.3:
$$(1, 2, 3)\begin{pmatrix} 1 & 2 & 3 \\ 4 & 5 & 6 \\ 7 & 8 & 9 \end{pmatrix} = (1+8+21,\ 2+10+24,\ 3+12+27) = (30, 36, 42)$$

Example 7.4: Consider the matrix $A = \begin{pmatrix} 2 & 1 \\ 2 & 3 \end{pmatrix}$. Then the vector $u = (2, -1)$ is a fixed point of A. For,
$$uA = (2, -1)\begin{pmatrix} 2 & 1 \\ 2 & 3 \end{pmatrix} = (2 \cdot 2 - 1 \cdot 2,\ 2 \cdot 1 - 1 \cdot 3) = (2, -1) = u$$
Thus by the above theorem, the vector $2u = (4, -2)$ is also a fixed point of A:
$$(4, -2)\begin{pmatrix} 2 & 1 \\ 2 & 3 \end{pmatrix} = (4 \cdot 2 - 2 \cdot 2,\ 4 \cdot 1 - 2 \cdot 3) = (4, -2)$$

PROBABILITY VECTORS, STOCHASTIC MATRICES

A vector $u = (u_1, u_2, \ldots, u_n)$ is called a *probability vector* if the components are non-negative and their sum is 1.

Example 7.5: Consider the following vectors:

$$u = (\tfrac{3}{4}, 0, -\tfrac{1}{4}, \tfrac{1}{2}), \quad v = (\tfrac{3}{4}, \tfrac{1}{2}, 0, \tfrac{1}{4}) \quad \text{and} \quad w = (\tfrac{1}{4}, \tfrac{1}{4}, 0, \tfrac{1}{2})$$

Then:

u is not a probability vector since its third component is negative;

v is not a probability vector since the sum of its components is greater than 1;

w is a probability vector.

Example 7.6: The nonzero vector $v = (2, 3, 5, 0, 1)$ is not a probability vector since the sum of its components is $2+3+5+0+1 = 11$. However, since the components of v are nonnegative, v has a unique scalar multiple λv which is a probability vector; it can be obtained from v by multiplying each component of v by the reciprocal of the sum of the components of v: $\frac{1}{11}v = (\frac{2}{11}, \frac{3}{11}, \frac{5}{11}, 0, \frac{1}{11})$.

A square matrix $P = (p_{ij})$ is called a *stochastic matrix* if each of its rows is a probability vector, i.e. if each entry of P is nonnegative and the sum of the entries in each row is 1.

Example 7.7: Consider the following matrices:

$$\begin{pmatrix} \frac{1}{3} & 0 & \frac{2}{3} \\ \frac{3}{4} & \frac{1}{2} & -\frac{1}{4} \\ \frac{1}{3} & \frac{1}{3} & \frac{1}{3} \end{pmatrix} \qquad \begin{pmatrix} \frac{1}{4} & \frac{3}{4} \\ \frac{1}{3} & \frac{1}{3} \end{pmatrix} \qquad \begin{pmatrix} 0 & 1 & 0 \\ \frac{1}{2} & \frac{1}{6} & \frac{1}{3} \\ \frac{1}{3} & \frac{2}{3} & 0 \end{pmatrix}$$

(i) (ii) (iii)

(i) is not a stochastic matrix since the entry in the second row and third column is negative;

(ii) is not a stochastic matrix since the sum of the entries in the second row is not 1;

(iii) is a stochastic matrix since each row is a probability vector.

We shall prove (see Problem 7.10)

Theorem 7.2: If A and B are stochastic matrices, then the product AB is a stochastic matrix. Therefore, in particular, all powers A^n are stochastic matrices.

REGULAR STOCHASTIC MATRICES

We now define an important class of stochastic matrices whose properties shall be investigated subsequently.

Definition: A stochastic matrix P is said to be *regular* if all the entries of some power P^m are positive.

Example 7.8: The stochastic matrix $A = \begin{pmatrix} 0 & 1 \\ \frac{1}{2} & \frac{1}{2} \end{pmatrix}$ is regular since

$$A^2 = \begin{pmatrix} 0 & 1 \\ \frac{1}{2} & \frac{1}{2} \end{pmatrix}\begin{pmatrix} 0 & 1 \\ \frac{1}{2} & \frac{1}{2} \end{pmatrix} = \begin{pmatrix} \frac{1}{2} & \frac{1}{2} \\ \frac{1}{4} & \frac{3}{4} \end{pmatrix}$$

is positive in every entry.

Example 7.9: Consider the stochastic matrix $A = \begin{pmatrix} 1 & 0 \\ \frac{1}{2} & \frac{1}{2} \end{pmatrix}$. Here

$$A^2 = \begin{pmatrix} 1 & 0 \\ \frac{3}{4} & \frac{1}{4} \end{pmatrix}, \quad A^3 = \begin{pmatrix} 1 & 0 \\ \frac{7}{8} & \frac{1}{8} \end{pmatrix}, \quad A^4 = \begin{pmatrix} 1 & 0 \\ \frac{15}{16} & \frac{1}{16} \end{pmatrix}$$

In fact every power A^m will have 1 and 0 in the first row; hence A is not regular.

FIXED POINTS AND REGULAR STOCHASTIC MATRICES

The fundamental property of regular stochastic matrices is contained in the following theorem whose proof lies beyond the scope of this text.

Theorem 7.3: Let P be a regular stochastic matrix. Then:

(i) P has a unique fixed probability vector t, and the components of t are all positive;

(ii) the sequence $P, P^2, P^3, \ldots$ of powers of P approaches the matrix T whose rows are each the fixed point t;

(iii) if p is any probability vector, then the sequence of vectors $pP, pP^2, pP^3, \ldots$ approaches the fixed point t.

Note: P^n approaches T means that each entry of P^n approaches the corresponding entry of T, and pP^n approaches t means that each component of pP^n approaches the corresponding component of t.

Example 7.10: Consider the regular stochastic matrix $P = \begin{pmatrix} 0 & 1 \\ \frac{1}{2} & \frac{1}{2} \end{pmatrix}$. We seek a probability vector with two components, which we can denote by $t = (x, 1-x)$, such that $tP = t$:

$$(x, 1-x)\begin{pmatrix} 0 & 1 \\ \frac{1}{2} & \frac{1}{2} \end{pmatrix} = (x, 1-x)$$

Multiplying the left side of the above matrix equation, we obtain

$$(\tfrac{1}{2} - \tfrac{1}{2}x, \tfrac{1}{2} + \tfrac{1}{2}x) = (x, 1-x) \quad \text{or} \quad \begin{cases} \frac{1}{2} - \frac{1}{2}x = x \\ \frac{1}{2} + \frac{1}{2}x = 1 - x \end{cases} \quad \text{or} \quad x = \tfrac{1}{3}$$

Thus $t = (\frac{1}{3}, 1 - \frac{1}{3}) = (\frac{1}{3}, \frac{2}{3})$ is the unique fixed probability vector of P. By Theorem 7.3, the sequence $P, P^2, P^3, \ldots$ approaches the matrix T whose rows are each the vector t:

$$T = \begin{pmatrix} \frac{1}{3} & \frac{2}{3} \\ \frac{1}{3} & \frac{2}{3} \end{pmatrix} = \begin{pmatrix} .33 & .67 \\ .33 & .67 \end{pmatrix}$$

We exhibit some of the powers of P to indicate the above result:

$$P^2 = \begin{pmatrix} \frac{1}{2} & \frac{1}{2} \\ \frac{1}{4} & \frac{3}{4} \end{pmatrix} = \begin{pmatrix} .50 & .50 \\ .25 & .75 \end{pmatrix}; \quad P^3 = \begin{pmatrix} \frac{1}{4} & \frac{3}{4} \\ \frac{3}{8} & \frac{5}{8} \end{pmatrix} = \begin{pmatrix} .25 & .75 \\ .37 & .63 \end{pmatrix}$$

$$P^4 = \begin{pmatrix} \frac{3}{8} & \frac{5}{8} \\ \frac{5}{16} & \frac{11}{16} \end{pmatrix} = \begin{pmatrix} .37 & .63 \\ .31 & .69 \end{pmatrix}; \quad P^5 = \begin{pmatrix} \frac{5}{16} & \frac{11}{16} \\ \frac{11}{32} & \frac{21}{32} \end{pmatrix} = \begin{pmatrix} .31 & .69 \\ .34 & .66 \end{pmatrix}$$

Example 7.11: Find the unique fixed probability vector of the regular stochastic matrix

$$P = \begin{pmatrix} 0 & 1 & 0 \\ 0 & 0 & 1 \\ \frac{1}{2} & \frac{1}{2} & 0 \end{pmatrix}$$

Method 1. We seek a probability vector with three components, which we can represent by $t = (x, y, 1-x-y)$, such that $tP = t$:

$$(x, y, 1-x-y)\begin{pmatrix} 0 & 1 & 0 \\ 0 & 0 & 1 \\ \frac{1}{2} & \frac{1}{2} & 0 \end{pmatrix} = (x, y, 1-x-y)$$

Multiplying the left side of the above matrix equation and then setting corresponding components equal to each other, we obtain the system

$$\begin{aligned} \tfrac{1}{2} - \tfrac{1}{2}x - \tfrac{1}{2}y &= x \\ x + \tfrac{1}{2} - \tfrac{1}{2}x - \tfrac{1}{2}y &= y \\ y &= 1 - x - y \end{aligned} \qquad \text{or} \qquad \begin{aligned} 3x + y &= 1 \\ x - 3y &= -1 \\ x + 2y &= 1 \end{aligned} \qquad \text{or} \qquad \begin{aligned} x &= \tfrac{1}{5} \\ y &= \tfrac{2}{5} \end{aligned}$$

Thus $t = (\frac{1}{5}, \frac{2}{5}, \frac{2}{5})$ is the unique fixed probability vector of P.

Method 2. We first seek any fixed vector $u = (x, y, z)$ of the matrix P:

$$(x, y, z)\begin{pmatrix} 0 & 1 & 0 \\ 0 & 0 & 1 \\ \frac{1}{2} & \frac{1}{2} & 0 \end{pmatrix} = (x, y, z) \qquad \text{or} \qquad \begin{cases} \frac{1}{2}z = x \\ x + \frac{1}{2}z = y \\ y = z \end{cases}$$

We know that the system has a nonzero solution; hence we can arbitrarily assign a value to one of the unknowns. Set $z = 2$. Then by the first equation $x = 1$, and by the third equation $y = 2$. Thus $u = (1, 2, 2)$ is a fixed point of P. But every multiple of u is a fixed point of P; hence multiply u by $\frac{1}{5}$ to obtain the required fixed probability vector $t = \frac{1}{5}u = (\frac{1}{5}, \frac{2}{5}, \frac{2}{5})$.

MARKOV CHAINS

We now consider a sequence of trials whose outcomes, say, $X_1, X_2, \ldots$, satisfy the following two properties:

(i) Each outcome belongs to a finite set of outcomes $\{a_1, a_2, \ldots, a_m\}$ called the *state space* of the system; if the outcome on the nth trial is a_i, then we say that the system is in state a_i at time n or at the nth step.

(ii) The outcome of any trial depends at most upon the outcome of the immediately preceding trial and not upon any other previous outcome; with each pair of states (a_i, a_j) there is given the probability p_{ij} that a_j occurs immediately after a_i occurs.

Such a stochastic process is called a (finite) *Markov chain.* The numbers p_{ij}, called the *transition probabilities,* can be arranged in a matrix

$$P = \begin{pmatrix} p_{11} & p_{12} & \cdots & p_{1m} \\ p_{21} & p_{22} & \cdots & p_{2m} \\ \cdots & \cdots & \cdots & \cdots \\ p_{m1} & p_{m2} & \cdots & p_{mm} \end{pmatrix}$$

called the *transition matrix.*

Thus with each state a_i there corresponds the ith row $(p_{i1}, p_{i2}, \ldots, p_{im})$ of the transition matrix P; if the system is in state a_i, then this row vector represents the probabilities of all the possible outcomes of the next trial and so it is a probability vector. Accordingly,

Theorem 7.4: The transition matrix P of a Markov chain is a stochastic matrix.

Example 7.12: A man either drives his car or takes a train to work each day. Suppose he never takes the train two days in a row; but if he drives to work, then the next day he is just as likely to drive again as he is to take the train.

The state space of the system is $\{t\,(\text{train}),\ d\,(\text{drive})\}$. This stochastic process is a Markov chain since the outcome on any day depends only on what happened the preceding day. The transition matrix of the Markov chain is

$$\begin{array}{c} \\ t \\ d \end{array} \begin{array}{c} t \quad d \\ \begin{pmatrix} 0 & 1 \\ \frac{1}{2} & \frac{1}{2} \end{pmatrix} \end{array}$$

The first row of the matrix corresponds to the fact that he never takes the train two days in a row and so he definitely will drive the day after he takes the train. The second row of the matrix corresponds to the fact that the day after he drives he will drive or take the train with equal probability.

Example 7.13: Three boys A, B and C are throwing a ball to each other. A always throws the ball to B and B always throws the ball to C; but C is just as likely to throw the ball to B as to A. Let X_n denote the nth person to be thrown the ball. The state space of the system is $\{A, B, C\}$. This is a Markov chain since the person throwing the ball is not influenced by those who previously had the ball. The transition matrix of the Markov chain is

$$\begin{array}{c} \\ A \\ B \\ C \end{array} \begin{array}{c} \begin{array}{ccc} A & B & C \end{array} \\ \begin{pmatrix} 0 & 1 & 0 \\ 0 & 0 & 1 \\ \frac{1}{2} & \frac{1}{2} & 0 \end{pmatrix} \end{array}$$

The first row of the matrix corresponds to the fact that A always throws the ball to B. The second row corresponds to the fact that B always throws the ball to C. The last row corresponds to the fact that C throws the ball to A or B with equal probability (and does not throw it to himself).

Example 7.14: A school contains 200 boys and 150 girls. One student is selected after another to take an eye examination. Let X_n denote the sex of the nth student who takes the examination. The state space of the stochastic process is $\{m\,(\text{male}), f\,(\text{female})\}$. However, this process is not a Markov chain since, for example, the probability that the third person is a girl depends not only on the outcome of the second trial but on both the first and second trials.

Example 7.15: (Random walk with reflecting barriers.) A man is at an integral point on the x-axis between the origin O and, say, the point 5. He takes a unit step to the right with probability p or to the left with probability $q = 1 - p$, unless he is at the origin where he takes a step to the right to 1 or at the point 5 where he takes a step to the left to 4. Let X_n denote his position after n steps. This is a Markov chain with state space $\{a_0, a_1, a_2, a_3, a_4, a_5\}$ where a_i means that the man is at the point i. The transition matrix is

$$P \quad = \quad \begin{array}{c} \\ a_0 \\ a_1 \\ a_2 \\ a_3 \\ a_4 \\ a_5 \end{array} \begin{array}{c} \begin{array}{cccccc} a_0 & a_1 & a_2 & a_3 & a_4 & a_5 \end{array} \\ \begin{pmatrix} 0 & 1 & 0 & 0 & 0 & 0 \\ q & 0 & p & 0 & 0 & 0 \\ 0 & q & 0 & p & 0 & 0 \\ 0 & 0 & q & 0 & p & 0 \\ 0 & 0 & 0 & q & 0 & p \\ 0 & 0 & 0 & 0 & 1 & 0 \end{pmatrix} \end{array}$$

Each row of the matrix, except the first and last, corresponds to the fact that the man moves from state a_i to state a_{i+1} with probability p or back to state a_{i-1} with probability $q = 1 - p$. The first row corresponds to the fact that the man must move from state a_0 to state a_1, and the last row that the man must move from state a_5 to state a_4.

HIGHER TRANSITION PROBABILITIES

The entry p_{ij} in the transition matrix P of a Markov chain is the probability that the system changes from the state a_i to the state a_j in one step: $a_i \to a_j$. Question: What is the probability, denoted by $p_{ij}^{(n)}$, that the system changes from the state a_i to the state a_j in exactly n steps:

$$a_i \to a_{k_1} \to a_{k_2} \to \cdots \to a_{k_{n-1}} \to a_j$$

The next theorem answers this question; here the $p_{ij}^{(n)}$ are arranged in a matrix $P^{(n)}$ called the *n-step transition matrix*:

Theorem 7.5: Let P be the transition matrix of a Markov chain process. Then the n-step transition matrix is equal to the nth power of P; that is, $P^{(n)} = P^n$.

Now suppose that, at some arbitrary time, the probability that the system is in state a_i is p_i; we denote these probabilities by the probability vector $p = (p_1, p_2, \ldots, p_m)$ which is called the *probability distribution* of the system at that time. In particular, we shall let

$$p^{(0)} = (p_1^{(0)}, p_2^{(0)}, \ldots, p_m^{(0)})$$

denote the *initial probability distribution*, i.e. the distribution when the process begins, and we shall let

$$p^{(n)} = (p_1^{(n)}, p_2^{(n)}, \ldots, p_m^{(n)})$$

denote the *n*th *step probability distribution*, i.e. the distribution after the first n steps. The following theorem applies.

Theorem 7.6: Let P be the transition matrix of a Markov chain process. If $p = (p_i)$ is the probability distribution of the system at some arbitrary time, then pP is the probability distribution of the system one step later and pP^n is the probability distribution of the system n steps later. In particular,

$$p^{(1)} = p^{(0)}P, \; p^{(2)} = p^{(1)}P, \; p^{(3)} = p^{(2)}P, \; \ldots, \; p^{(n)} = p^{(0)}P^n$$

Example 7.16: Consider the Markov chain of Example 7.12 whose transition matrix is

$$P = \begin{array}{c} \\ t \\ d \end{array} \begin{array}{c} \begin{array}{cc} t & d \end{array} \\ \begin{pmatrix} 0 & 1 \\ \frac{1}{2} & \frac{1}{2} \end{pmatrix} \end{array}$$

Here t is the state of taking a train to work and d of driving to work. By Example 7.8,

$$P^4 = P^2 \cdot P^2 = \begin{pmatrix} \frac{1}{2} & \frac{1}{2} \\ \frac{1}{4} & \frac{3}{4} \end{pmatrix} \begin{pmatrix} \frac{1}{2} & \frac{1}{2} \\ \frac{1}{4} & \frac{3}{4} \end{pmatrix} = \begin{pmatrix} \frac{3}{8} & \frac{5}{8} \\ \frac{5}{16} & \frac{11}{16} \end{pmatrix}$$

Thus the probability that the system changes from, say, state t to state d in exactly 4 steps is $\frac{5}{8}$, i.e. $p_{td}^{(4)} = \frac{5}{8}$. Similarly, $p_{tt}^{(4)} = \frac{3}{8}$, $p_{dt}^{(4)} = \frac{5}{16}$ and $p_{dd}^{(4)} = \frac{11}{16}$.

Now suppose that on the first day of work, the man tossed a fair die and drove to work if and only if a 6 appeared. In other words, $p^{(0)} = (\frac{5}{6}, \frac{1}{6})$ is the initial probability distribution. Then

$$p^{(4)} = p^{(0)}P^4 = (\tfrac{5}{6}, \tfrac{1}{6}) \begin{pmatrix} \frac{3}{8} & \frac{5}{8} \\ \frac{5}{16} & \frac{11}{16} \end{pmatrix} = (\tfrac{35}{96}, \tfrac{61}{96})$$

is the probability distribution after 4 days, i.e. $p_t^{(4)} = \frac{35}{96}$ and $p_d^{(4)} = \frac{61}{96}$.

Example 7.17: Consider the Markov chain of Example 7.13 whose transition matrix is

$$P = \begin{array}{c} \\ A \\ B \\ C \end{array} \begin{array}{c} \begin{array}{ccc} A & B & C \end{array} \\ \begin{pmatrix} 0 & 1 & 0 \\ 0 & 0 & 1 \\ \frac{1}{2} & \frac{1}{2} & 0 \end{pmatrix} \end{array}$$

Suppose C was the first person with the ball, i.e. suppose $p^{(0)} = (0,0,1)$ is the initial probability distribution. Then

$$p^{(1)} = p^{(0)}P = (0,0,1)\begin{pmatrix} 0 & 1 & 0 \\ 0 & 0 & 1 \\ \frac{1}{2} & \frac{1}{2} & 0 \end{pmatrix} = (\tfrac{1}{2}, \tfrac{1}{2}, 0)$$

$$p^{(2)} = p^{(1)}P = (\tfrac{1}{2}, \tfrac{1}{2}, 0)\begin{pmatrix} 0 & 1 & 0 \\ 0 & 0 & 1 \\ \frac{1}{2} & \frac{1}{2} & 0 \end{pmatrix} = (0, \tfrac{1}{2}, \tfrac{1}{2})$$

$$p^{(3)} = p^{(2)}P = (0, \tfrac{1}{2}, \tfrac{1}{2})\begin{pmatrix} 0 & 1 & 0 \\ 0 & 0 & 1 \\ \frac{1}{2} & \frac{1}{2} & 0 \end{pmatrix} = (\tfrac{1}{4}, \tfrac{1}{4}, \tfrac{1}{2})$$

Thus, after three throws, the probability that A has the ball is $\frac{1}{4}$, that B has the ball is $\frac{1}{4}$ and that C has the ball is $\frac{1}{2}$: $p_A^{(3)} = \frac{1}{4}$, $p_B^{(3)} = \frac{1}{4}$ and $p_C^{(3)} = \frac{1}{2}$.

Example 7.18: Consider the random walk problem of Example 7.15. Suppose the man began at the point 2; find the probability distribution after 3 steps and after 4 steps, i.e. $p^{(3)}$ and $p^{(4)}$.

Now $p^{(0)} = (0,0,1,0,0,0)$ is the initial probability distribution. Then

$$p^{(1)} = p^{(0)}P = (0, q, 0, p, 0, 0)$$

$$p^{(2)} = p^{(1)}P = (q^2, 0, 2pq, 0, p^2, 0)$$

$$p^{(3)} = p^{(2)}P = (0, q^2 + 2pq^2, 0, 3p^2q, 0, p^3)$$

$$p^{(4)} = p^{(3)}P = (q^3 + 2pq^3, 0, pq^2 + 5p^2q^2, 0, 3p^3q + p^3, 0)$$

Thus after 4 steps he is at, say, the origin with probability $q^3 + 2pq^3$.

STATIONARY DISTRIBUTION OF REGULAR MARKOV CHAINS

Suppose that a Markov chain is regular, i.e. that its transition matrix P is regular. By Theorem 7.3 the sequence of n-step transition matrices P^n approaches the matrix T whose rows are each the unique fixed probability vector t of P; hence the probability $p_{ij}^{(n)}$ that a_j occurs for sufficiently large n is independent of the original state a_i and it approaches the component t_j of t. In other words,

Theorem 7.7: Let the transition matrix P of a Markov chain be regular. Then, in the long run, the probability that any state a_j occurs is approximately equal to the component t_j of the unique fixed probability vector t of P.

Thus we see that the effect of the initial state or the initial probability distribution of the process wears off as the number of steps of the process increase. Furthermore, every sequence of probability distributions approaches the fixed probability vector t of P, called the *stationary distribution* of the Markov chain.

Example 7.19: Consider the Markov chain process of Example 7.12 whose transition matrix is

$$P = \begin{matrix} & \begin{matrix} t & d \end{matrix} \\ \begin{matrix} t \\ d \end{matrix} & \begin{pmatrix} 0 & 1 \\ \frac{1}{2} & \frac{1}{2} \end{pmatrix} \end{matrix}$$

By Example 7.10, the unique fixed probability vector of the above matrix is $(\frac{1}{3}, \frac{2}{3})$. Thus, in the long run, the man will take the train to work $\frac{1}{3}$ of the time, and drive to work the other $\frac{2}{3}$ of the time.

Example 7.20: Consider the Markov chain process of Example 7.13 whose transition matrix is

$$P = \begin{array}{c} \\ A \\ B \\ C \end{array} \begin{array}{c} \begin{array}{ccc} A & B & C \end{array} \\ \begin{pmatrix} 0 & 1 & 0 \\ 0 & 0 & 1 \\ \frac{1}{2} & \frac{1}{2} & 0 \end{pmatrix} \end{array}$$

By Example 7.11, the unique fixed probability vector of the above matrix is $(\frac{1}{5}, \frac{2}{5}, \frac{2}{5})$. Thus, in the long run, A will be thrown the ball 20% of the time, and B and C 40% of the time.

ABSORBING STATES

A state a_i of a Markov chain is called *absorbing* if the system remains in the state a_i once it enters there. Thus a state a_i is absorbing if and only if the ith row of the transition matrix P has a 1 on the main diagonal and zeros everywhere else. (The *main diagonal* of an n-square matrix $A = (a_{ij})$ consists of the entries $a_{11}, a_{22}, \ldots, a_{nn}$.)

Example 7.21: Suppose the following matrix is the transition matrix of a Markov chain:

$$P = \begin{array}{c} \\ a_1 \\ a_2 \\ a_3 \\ a_4 \\ a_5 \end{array} \begin{array}{c} \begin{array}{ccccc} a_1 & a_2 & a_3 & a_4 & a_5 \end{array} \\ \begin{pmatrix} \frac{1}{4} & 0 & \frac{1}{4} & \frac{1}{4} & \frac{1}{4} \\ 0 & 1 & 0 & 0 & 0 \\ \frac{1}{2} & 0 & \frac{1}{4} & \frac{1}{4} & 0 \\ 0 & 1 & 0 & 0 & 0 \\ 0 & 0 & 0 & 0 & 1 \end{pmatrix} \end{array}$$

The states a_2 and a_5 are each absorbing, since each of the second and fifth rows has a 1 on the main diagonal.

Example 7.22: (Random walk with absorbing barriers.) Consider the random walk problem of Example 7.15, except now we assume that the man remains at either endpoint whenever he reaches there. This is also a Markov chain and the transition matrix is given by

$$P = \begin{array}{c} \\ a_0 \\ a_1 \\ a_2 \\ a_3 \\ a_4 \\ a_5 \end{array} \begin{array}{c} \begin{array}{cccccc} a_0 & a_1 & a_2 & a_3 & a_4 & a_5 \end{array} \\ \begin{pmatrix} 1 & 0 & 0 & 0 & 0 & 0 \\ q & 0 & p & 0 & 0 & 0 \\ 0 & q & 0 & p & 0 & 0 \\ 0 & 0 & q & 0 & p & 0 \\ 0 & 0 & 0 & q & 0 & p \\ 0 & 0 & 0 & 0 & 0 & 1 \end{pmatrix} \end{array}$$

We call this process a random walk with absorbing barriers, since the a_0 and a_5 are absorbing states. In this case, $p_0^{(n)}$ denotes the probability that the man reaches the state a_0 on or before the nth step. Similarly, $p_5^{(n)}$ denotes the probability that he reaches the state a_5 on or before the nth step.

Example 7.23: A player has, say, x dollars. He bets one dollar at a time and wins with probability p and loses with probability $q = 1 - p$. The game ends when he loses all his money, i.e. has 0 dollars, or when he wins $N - x$ dollars, i.e. has N dollars. This game is identical to the random walk of the preceding example except that here the absorbing barriers are at 0 and N.

Example 7.24: A man tosses a fair coin until 3 heads occur in a row. Let $X_n = k$ if, at the nth trial, the last tail occurred at the $(n-k)$-th trial, i.e. X_n denotes the longest string of heads ending at the nth trial. This is a Markov chain process with state space $\{a_0, a_1, a_2, a_3\}$, where a_i means the string of heads has length i. The transition matrix is

$$\begin{array}{c} \\ a_0 \\ a_1 \\ a_2 \\ a_3 \end{array} \begin{array}{c} \begin{array}{cccc} a_0 & a_1 & a_2 & a_3 \end{array} \\ \begin{pmatrix} \frac{1}{2} & \frac{1}{2} & 0 & 0 \\ \frac{1}{2} & 0 & \frac{1}{2} & 0 \\ \frac{1}{2} & 0 & 0 & \frac{1}{2} \\ 0 & 0 & 0 & 1 \end{pmatrix} \end{array}$$

Each row, except the last, corresponds to the fact that a string of heads is either broken if a tail occurs or is extended by one if a head occurs. The last line corresponds to the fact that the game ends if three heads are tossed in a row. Note that a_3 is an absorbing state.

Let a_i be an absorbing state of a Markov chain with transition matrix P. Then, for $j \neq i$, the n-step transition probability $p_{ij}^{(n)} = 0$ for every n. Accordingly, every power of P has a zero entry and so P is not regular. Thus:

Theorem 7.8: If a stochastic matrix P has a 1 on the main diagonal, then P is not regular (unless P is a 1×1 matrix).

Solved Problems

MATRIX MULTIPLICATION

7.1. Let $u = (1, -2, 4)$ and $A = \begin{pmatrix} 1 & 3 & -1 \\ 0 & 2 & 5 \\ 4 & 1 & 6 \end{pmatrix}$. Find uA.

The product of the vector u with 3 components by the 3×3 matrix A is again a vector with 3 components. To obtain the first component of uA, multiply the elements of u by the corresponding elements of the first column of A and then add:

$$(1, -2, 4)\begin{pmatrix} 1 & 3 & -1 \\ 0 & 2 & 5 \\ 4 & 1 & 6 \end{pmatrix} = (1 \cdot 1 + (-2) \cdot 0 + 4 \cdot 4, \quad , \quad) = (17, \ , \ )$$

To obtain the second component of uA, multiply the elements of u by the corresponding elements of the second column of A and then add:

$$(1, -2, 4)\begin{pmatrix} 1 & 3 & -1 \\ 0 & 2 & 5 \\ 4 & 1 & 6 \end{pmatrix} = (17, 1 \cdot 3 + (-2) \cdot 2 + 4 \cdot 1, \quad) = (17, 3, \ )$$

To obtain the third component of uA, multiply the elements of u by the corresponding elements of the third column of A and then add:

$$(1, -2, 4)\begin{pmatrix} 1 & 3 & -1 \\ 0 & 2 & 5 \\ 4 & 1 & 6 \end{pmatrix} = (17, 3, 1 \cdot (-1) + (-2) \cdot 5 + 4 \cdot 6) = (17, 3, 13)$$

That is,

$$uA = (17, 3, 13)$$

7.2. Let $A = \begin{pmatrix} 1 & 3 \\ 2 & -1 \end{pmatrix}$ and $B = \begin{pmatrix} 2 & 0 & -4 \\ 3 & -2 & 6 \end{pmatrix}$. Find (i) AB, (ii) BA.

(i) Since A is 2×2 and B is 2×3, the product AB is a 2×3 matrix. To obtain the first row of AB, multiply the elements of the first row (1, 3) of A by the corresponding elements of each of the columns $\begin{pmatrix} 2 \\ 3 \end{pmatrix}$, $\begin{pmatrix} 0 \\ -2 \end{pmatrix}$ and $\begin{pmatrix} -4 \\ 6 \end{pmatrix}$ of B and then add:

$$\begin{pmatrix} 1 & 3 \\ 2 & -1 \end{pmatrix}\begin{pmatrix} 2 & 0 & -4 \\ 3 & -2 & 6 \end{pmatrix}$$

$$= \begin{pmatrix} 1 \cdot 2 + 3 \cdot 3 & 1 \cdot 0 + 3 \cdot (-2) & 1 \cdot (-4) + 3 \cdot 6 \\ & & \end{pmatrix} = \begin{pmatrix} 11 & -6 & 14 \\ & & \end{pmatrix}$$

To obtain the second row of AB, multiply the elements of the second row $(2, -1)$ of A by the corresponding elements of each of the columns of B and then add:

$$\begin{pmatrix} 1 & 3 \\ 2 & -1 \end{pmatrix}\begin{pmatrix} 2 & 0 & -4 \\ 3 & -2 & 6 \end{pmatrix}$$

$$= \begin{pmatrix} 11 & -6 & 14 \\ 2 \cdot 2 + (-1) \cdot 3 & 2 \cdot 0 + (-1) \cdot (-2) & 2 \cdot (-4) + (-1) \cdot 6 \end{pmatrix} = \begin{pmatrix} 11 & -6 & 14 \\ 1 & 2 & -14 \end{pmatrix}$$

Thus $$AB = \begin{pmatrix} 11 & -6 & 14 \\ 1 & 2 & -14 \end{pmatrix}$$

(ii) Note B is 2×3 and A is 2×2. Since the "inner numbers" 3 and 2 are not equal, i.e. the number of columns of B is not equal to the number of rows of A, the product BA is not defined.

7.3. Let $A = \begin{pmatrix} 1 & 2 \\ 4 & -3 \end{pmatrix}$. Find (i) A^2, (ii) A^3.

(i) $$A^2 = AA = \begin{pmatrix} 1 & 2 \\ 4 & -3 \end{pmatrix}\begin{pmatrix} 1 & 2 \\ 4 & -3 \end{pmatrix}$$

$$= \begin{pmatrix} 1 \cdot 1 + 2 \cdot 4 & 1 \cdot 2 + 2 \cdot (-3) \\ 4 \cdot 1 + (-3) \cdot 4 & 4 \cdot 2 + (-3) \cdot (-3) \end{pmatrix} = \begin{pmatrix} 9 & -4 \\ -8 & 17 \end{pmatrix}$$

(ii) $$A^3 = AA^2 = \begin{pmatrix} 1 & 2 \\ 4 & -3 \end{pmatrix}\begin{pmatrix} 9 & -4 \\ -8 & 17 \end{pmatrix}$$

$$= \begin{pmatrix} 1 \cdot 9 + 2 \cdot (-8) & 1 \cdot (-4) + 2 \cdot 17 \\ 4 \cdot 9 + (-3) \cdot (-8) & 4 \cdot (-4) + (-3) \cdot 17 \end{pmatrix} = \begin{pmatrix} -7 & 30 \\ 60 & -67 \end{pmatrix}$$

PROBABILITY VECTORS AND STOCHASTIC MATRICES

7.4. Which vectors are probability vectors?

(i) $u = (\frac{1}{3}, 0, -\frac{1}{6}, \frac{1}{2}, \frac{1}{3})$, (ii) $v = (\frac{1}{3}, 0, \frac{1}{6}, \frac{1}{2}, \frac{1}{3})$, (iii) $w = (\frac{1}{3}, 0, 0, \frac{1}{6}, \frac{1}{2})$.

A vector is a probability vector if its components are nonnegative and their sum is 1.

(i) u is not a probability vector since its third component is negative.

(ii) v is not a probability vector since the sum of the components is greater than 1.

(iii) w is a probability vector since the components are nonnegative and their sum is 1.

7.5. Multiply each vector by the appropriate scalar to form a probability vector:

(i) (2, 1, 0, 2, 3), (ii) (4, 0, 1, 2, 0, 5), (iii) (3, 0, −2, 1), (iv) (0, 0, 0, 0, 0).

(i) The sum of the components is $2 + 1 + 0 + 3 + 2 = 8$; hence multiply the vector, i.e. each component, by $\frac{1}{8}$ to obtain the probability vector $(\frac{1}{4}, \frac{1}{8}, 0, \frac{1}{4}, \frac{3}{8})$.

(ii) The sum of the components is $4+0+1+2+0+5 = 12$; hence multiply the vector, i.e. each component, by $\frac{1}{12}$ to obtain the probability vector $(\frac{1}{3}, 0, \frac{1}{12}, \frac{1}{6}, 0, \frac{5}{12})$.

(iii) The first component is positive and the third is negative; hence it is impossible to multiply the vector by a scalar to form a vector with nonnegative components. Thus no scalar multiple of the vector is a probability vector.

(iv) Every scalar multiple of the zero vector is the zero vector whose components add to 0. Thus no multiple of the zero vector is a probability vector.

7.6. Find a multiple of each vector which is a probability vector: (i) $(\frac{1}{2}, \frac{2}{3}, 0, 2, \frac{5}{6})$, (ii) $(0, \frac{2}{3}, 1, \frac{3}{5}, \frac{5}{6})$.

In each case, first multiply each vector by a scalar so that the fractions are eliminated.

(i) First multiply the vector by 6 to obtain $(3, 4, 0, 12, 5)$. Then multiply by $1/(3+4+0+12+5) = \frac{1}{24}$ to obtain $(\frac{1}{8}, \frac{1}{6}, 0, \frac{1}{2}, \frac{5}{24})$ which is a probability vector.

(ii) First multiply the vector by 30 to obtain $(0, 20, 30, 18, 25)$. Then multiply by $1/(0+20+30+18+25) = \frac{1}{93}$ to obtain $(0, \frac{20}{93}, \frac{30}{93}, \frac{18}{93}, \frac{25}{93})$ which is a probability vector.

7.7. Which of the following matrices are stochastic matrices?

$$\text{(i) } A = \begin{pmatrix} \frac{1}{3} & \frac{1}{3} & \frac{1}{3} \\ \frac{1}{2} & 0 & \frac{1}{2} \end{pmatrix} \quad \text{(ii) } B = \begin{pmatrix} \frac{15}{16} & \frac{1}{16} \\ \frac{2}{3} & \frac{2}{3} \end{pmatrix} \quad \text{(iii) } C = \begin{pmatrix} 1 & 0 \\ \frac{1}{2} & \frac{1}{2} \end{pmatrix} \quad \text{(iv) } D = \begin{pmatrix} \frac{1}{2} & -\frac{1}{2} \\ \frac{1}{4} & \frac{3}{4} \end{pmatrix}.$$

(i) A is not a stochastic matrix since it is not a square matrix.

(ii) B is not a stochastic matrix since the sum of the components in the last row is greater than 1.

(iii) C is a stochastic matrix.

(iv) D is not a stochastic matrix since the entry in the first row, second column is negative.

7.8. Let $A = \begin{pmatrix} a_1 & b_1 & c_1 \\ a_2 & b_2 & c_2 \\ a_3 & b_3 & c_3 \end{pmatrix}$ be a stochastic matrix and let $u = (u_1, u_2, u_3)$ be a probability vector. Show that uA is also a probability vector.

$$uA = (u_1, u_2, u_3) \begin{pmatrix} a_1 & b_1 & c_1 \\ a_2 & b_2 & c_2 \\ a_3 & b_3 & c_3 \end{pmatrix} = (u_1a_1 + u_2a_2 + u_3a_3,\ u_1b_1 + u_2b_2 + u_3b_3,\ u_1c_1 + u_2c_2 + u_3c_3)$$

Since the u_i, a_i, b_i and c_i are nonnegative and since the products and sums of nonnegative numbers are nonnegative, the components of uA are nonnegative as required. Thus we only need to show that the sum of the components of uA is 1. Here we use the fact that $u_1 + u_2 + u_3$, $a_1 + b_1 + c_1$, $a_2 + b_2 + c_2$ and $a_3 + b_3 + c_3$ are each 1:

$$\begin{aligned} & u_1a_1 + u_2a_2 + u_3a_3 + u_1b_1 + u_2b_2 + u_3b_3 + u_1c_1 + u_2c_2 + u_3c_3 \\ &= u_1(a_1+b_1+c_1) + u_2(a_2+b_2+c_2) + u_3(a_3+b_3+c_3) \\ &= u_1 \cdot 1 + u_2 \cdot 1 + u_3 \cdot 1 = u_1 + u_2 + u_3 = 1 \end{aligned}$$

7.9. Prove: If $A = (a_{ij})$ is a stochastic matrix of order n and $u = (u_1, u_2, \ldots, u_n)$ is a probability vector, then uA is also a probability vector.

The proof is similar to that of the preceding problem for the case $n = 3$:

$$uA \;=\; (u_1, u_2, \ldots, u_n)\begin{pmatrix} a_{11} & a_{12} & \cdots & a_{1n} \\ a_{21} & a_{22} & \cdots & a_{2n} \\ \cdots & \cdots & \cdots & \cdots \\ a_{n1} & a_{n2} & \cdots & a_{nn} \end{pmatrix}$$

$$= \;(u_1a_{11} + u_2a_{21} + \cdots + u_na_{n1},\; u_1a_{12} + u_2a_{22} + \cdots + u_na_{n2},\; \ldots,\; u_1a_{1n} + u_2a_{2n} + \cdots + u_na_{nn})$$

Since the u_i and a_{ij} are nonnegative, the components of uA are also nonnegative. Thus we only need to show that the sum of the components of uA is 1:

$$u_1a_{11} + u_2a_{21} + \cdots + u_na_{n1} + u_1a_{12} + u_2a_{22} + \cdots + u_na_{n2} + \cdots + u_1a_{1n} + u_2a_{2n} + \cdots + u_na_{nn}$$

$$= \;u_1(a_{11} + a_{12} + \cdots + a_{1n}) + u_2(a_{21} + a_{22} + \cdots + a_{2n}) + \cdots + u_n(a_{n1} + a_{n2} + \cdots + a_{nn})$$

$$= \;u_1 \cdot 1 + u_2 \cdot 1 + \cdots + u_n \cdot 1 \;=\; u_1 + u_2 + \cdots + u_n \;=\; 1$$

7.10. Prove Theorem 7.2: If A and B are stochastic matrices, then the product AB is a stochastic matrix. Therefore, in particular, all powers A^n are stochastic matrices.

The ith-row s_i of the product matrix AB is obtained by multiplying the ith-row r_i of A by the matrix B: $s_i = r_iB$. Since each r_i is a probability vector and B is a stochastic matrix, by the preceding problem, s_i is also a probability vector. Hence AB is a stochastic matrix.

7.11. Prove: Let $p = (p_1, p_2, \ldots, p_m)$ be a probability vector, and let T be a matrix whose rows are each the same vector $t = (t_1, t_2, \ldots, t_m)$. Then $pT = t$.

Using the fact that $p_1 + p_2 + \cdots + p_m = 1$, we have

$$pT \;=\; (p_1, p_2, \ldots, p_m)\begin{pmatrix} t_1 & t_2 & \cdots & t_m \\ t_1 & t_2 & \cdots & t_m \\ \cdots & \cdots & \cdots & \cdots \\ t_1 & t_2 & \cdots & t_m \end{pmatrix}$$

$$= \;(p_1t_1 + p_2t_1 + \cdots + p_mt_1,\; p_1t_2 + p_2t_2 + \cdots + p_mt_2,\; \ldots,\; p_1t_m + p_2t_m + \cdots + p_mt_m)$$

$$= \;((p_1 + p_2 + \cdots + p_m)t_1,\; (p_1 + p_2 + \cdots + p_m)t_2,\; \ldots,\; (p_1 + p_2 + \cdots + p_m)t_m)$$

$$= \;(1 \cdot t_1, 1 \cdot t_2, \ldots, 1 \cdot t_m) \;=\; (t_1, t_2, \ldots, t_m) \;=\; t$$

REGULAR STOCHASTIC MATRICES AND FIXED PROBABILITY VECTORS

7.12. Find the unique fixed probability vector of the regular stochastic matrix $A = \begin{pmatrix} \frac{3}{4} & \frac{1}{4} \\ \frac{1}{2} & \frac{1}{2} \end{pmatrix}$. What matrix does A^n approach?

We seek a probability vector $t = (x, 1-x)$ such that $tA = t$:

$$(x, 1-x)\begin{pmatrix} \frac{3}{4} & \frac{1}{4} \\ \frac{1}{2} & \frac{1}{2} \end{pmatrix} \;=\; (x, 1-x)$$

Multiply the left side of the above matrix equation and then set corresponding components equal to each other to obtain the two equations

$$\tfrac{3}{4}x + \tfrac{1}{2} - \tfrac{1}{2}x \;=\; x, \qquad \tfrac{1}{4}x + \tfrac{1}{2} - \tfrac{1}{2}x \;=\; 1 - x$$

Solve either equation to obtain $x = \frac{2}{3}$. Thus $t = (\frac{2}{3}, \frac{1}{3})$ is the required probability vector.

Check the answer by computing the product tA:

$$(\tfrac{2}{3}, \tfrac{1}{3})\begin{pmatrix} \frac{3}{4} & \frac{1}{4} \\ \frac{1}{2} & \frac{1}{2} \end{pmatrix} \;=\; (\tfrac{1}{2} + \tfrac{1}{6}, \tfrac{1}{6} + \tfrac{1}{6}) \;=\; (\tfrac{2}{3}, \tfrac{1}{3})$$

The answer checks since $tA = t$.

The matrix A^n approaches the matrix T whose rows are each the fixed point t: $T = \begin{pmatrix} \frac{2}{3} & \frac{1}{3} \\ \frac{2}{3} & \frac{1}{3} \end{pmatrix}$.

7.13. (i) Show that the vector $u = (b, a)$ is a fixed point of the general 2×2 stochastic matrix $P = \begin{pmatrix} 1-a & a \\ b & 1-b \end{pmatrix}$.

(ii) Use the result of (i) to find the unique fixed probability vector of each of the following matrices:

$$A = \begin{pmatrix} \frac{1}{3} & \frac{2}{3} \\ 1 & 0 \end{pmatrix} \qquad B = \begin{pmatrix} \frac{1}{2} & \frac{1}{2} \\ \frac{2}{3} & \frac{1}{3} \end{pmatrix} \qquad C = \begin{pmatrix} .7 & .3 \\ .8 & .2 \end{pmatrix}$$

(i) $$uP \;=\; (b, a)\begin{pmatrix} 1-a & a \\ b & 1-b \end{pmatrix} \;=\; (b - ab + ab,\ ab + a - ab) \;=\; (b, a) \;=\; u.$$

(ii) By (i), $u = (1, \frac{2}{3})$ is a fixed point of A. Multiply u by 3 to obtain the fixed point $(3, 2)$ of A which has no fractions. Then multiply $(3, 2)$ by $1/(3+2) = \frac{1}{5}$ to obtain the required unique fixed probability vector $(\frac{3}{5}, \frac{2}{5})$.

By (i), $u = (\frac{2}{3}, \frac{1}{2})$ is a fixed point of B. Multiply u by 6 to obtain the fixed point $(4, 3)$, and then multiply by $1/(4+3) = \frac{1}{7}$ to obtain the required unique fixed probability vector $(\frac{4}{7}, \frac{3}{7})$.

By (i), $u = (.8, .3)$ is a fixed point of C. Hence $(8, 3)$ and the probability vector $(\frac{8}{11}, \frac{3}{11})$ are also fixed points of C.

7.14. Find the unique fixed probability vector of the regular stochastic matrix

$$P \;=\; \begin{pmatrix} \frac{1}{2} & \frac{1}{4} & \frac{1}{4} \\ \frac{1}{2} & 0 & \frac{1}{2} \\ 0 & 1 & 0 \end{pmatrix}$$

Method 1. We seek a probability vector $t = (x, y, 1-x-y)$ such that $tP = t$:

$$(x, y, 1-x-y)\begin{pmatrix} \frac{1}{2} & \frac{1}{4} & \frac{1}{4} \\ \frac{1}{2} & 0 & \frac{1}{2} \\ 0 & 1 & 0 \end{pmatrix} \;=\; (x, y, 1-x-y)$$

Multiply the left side of the above matrix equation and then set corresponding components equal to each other to obtain the system of three equations

$$\begin{cases} \frac{1}{2}x + \frac{1}{2}y \;=\; x \\ \frac{1}{4}x + 1 - x - y \;=\; y \\ \frac{1}{4}x + \frac{1}{2}y \;=\; 1 - x - y \end{cases} \qquad \text{or} \qquad \begin{cases} x - y \;=\; 0 \\ 3x + 8y \;=\; 4 \\ 5x + 6y \;=\; 4 \end{cases}$$

Choose any two of the equations and solve for x and y to obtain $x = \frac{4}{11}$ and $y = \frac{4}{11}$. Check the solution by substituting for x and y into the third equation. Since $1 - x - y = \frac{3}{11}$, the required fixed probability vector is $t = (\frac{4}{11}, \frac{4}{11}, \frac{3}{11})$.

Method 2. We seek any fixed vector $u = (x, y, z)$ of the matrix P:

$$(x, y, z)\begin{pmatrix} \frac{1}{2} & \frac{1}{4} & \frac{1}{4} \\ \frac{1}{2} & 0 & \frac{1}{2} \\ 0 & 1 & 0 \end{pmatrix} \;=\; (x, y, z)$$

Multiply the left side of the above matrix equation and set corresponding components equal to each other to obtain the system of three equations

$$\begin{cases} \frac{1}{2}x + \frac{1}{2}y \;=\; x \\ \frac{1}{4}x + z \;=\; y \\ \frac{1}{4}x + \frac{1}{2}y \;=\; z \end{cases} \qquad \text{or} \qquad \begin{cases} x - y \;=\; 0 \\ x - 4y + 4z \;=\; 0 \\ x + 2y - 4z \;=\; 0 \end{cases}$$

We know that the system has a nonzero solution; hence we can arbitrarily assign a value to one of the unknowns. Set $y = 4$. Then by the first equation $x = 4$, and by the third equation $z = 3$. Thus $u = (4, 4, 3)$ is a fixed point of P. Multiply u by $1/(4+4+3) = \frac{1}{11}$ to obtain $t = \frac{1}{11}u = (\frac{4}{11}, \frac{4}{11}, \frac{3}{11})$ which is a probability vector and is also a fixed point of P.

7.15. Find the unique fixed probability vector of the regular stochastic matrix

$$P = \begin{pmatrix} 0 & 1 & 0 \\ \frac{1}{6} & \frac{1}{2} & \frac{1}{3} \\ 0 & \frac{2}{3} & \frac{1}{3} \end{pmatrix}$$

What matrix does P^n approach?

We first seek any fixed vector $u = (x, y, z)$ of the matrix P:

$$(x, y, z) \begin{pmatrix} 0 & 1 & 0 \\ \frac{1}{6} & \frac{1}{2} & \frac{1}{3} \\ 0 & \frac{2}{3} & \frac{1}{3} \end{pmatrix} = (x, y, z)$$

Multiply the left side of the above matrix equation and set corresponding components equal to each other to obtain the system of three equations

$$\begin{cases} \frac{1}{6}y = x \\ x + \frac{1}{2}y + \frac{2}{3}z = y \\ \frac{1}{3}y + \frac{1}{3}z = z \end{cases} \quad \text{or} \quad \begin{cases} y = 6x \\ 6x + 3y + 4z = 6y \\ y + z = 3z \end{cases} \quad \text{or} \quad \begin{cases} y = 6x \\ 6x + 4z = 3y \\ y = 2z \end{cases}$$

We know that the system has a nonzero solution; hence we can arbitrarily assign a value to one of the unknowns. Set $x = 1$. Then by the first equation $y = 6$, and by the last equation $z = 3$. Thus $u = (1, 6, 3)$ is a fixed point of P. Since $1 + 6 + 3 = 10$, the vector $t = (\frac{1}{10}, \frac{6}{10}, \frac{3}{10})$ is the required unique fixed probability vector of P.

P^n approaches the matrix T whose rows are each the fixed point t: $T = \begin{pmatrix} \frac{1}{10} & \frac{6}{10} & \frac{3}{10} \\ \frac{1}{10} & \frac{6}{10} & \frac{3}{10} \\ \frac{1}{10} & \frac{6}{10} & \frac{3}{10} \end{pmatrix}$.

7.16. If $t = (\frac{1}{4}, 0, \frac{1}{2}, \frac{1}{4}, 0)$ is a fixed point of a stochastic matrix P, why is P not regular?

If P is regular then, by Theorem 7.3, P has a unique fixed probability vector, and the components of the vector are positive. Since the components of the given fixed probability vector are not all positive, P cannot be regular.

7.17. Which of the following stochastic matrices are regular?

(i) $A = \begin{pmatrix} \frac{1}{2} & \frac{1}{2} \\ 0 & 1 \end{pmatrix}$ (ii) $B = \begin{pmatrix} 0 & 1 \\ 1 & 0 \end{pmatrix}$ (iii) $C = \begin{pmatrix} \frac{1}{2} & \frac{1}{4} & \frac{1}{4} \\ 0 & 1 & 0 \\ \frac{1}{2} & \frac{1}{2} & 0 \end{pmatrix}$ (iv) $D = \begin{pmatrix} 0 & 0 & 1 \\ \frac{1}{2} & \frac{1}{4} & \frac{1}{4} \\ 0 & 1 & 0 \end{pmatrix}$

Recall that a stochastic matrix is regular if a power of the matrix has only positive entries.

(i) A is not regular since there is a 1 on the main diagonal (in the second row).

(ii) $B^2 = \begin{pmatrix} 0 & 1 \\ 1 & 0 \end{pmatrix}\begin{pmatrix} 0 & 1 \\ 1 & 0 \end{pmatrix} = \begin{pmatrix} 1 & 0 \\ 0 & 1 \end{pmatrix} =$ the identity matrix I

$B^3 = \begin{pmatrix} 1 & 0 \\ 0 & 1 \end{pmatrix}\begin{pmatrix} 0 & 1 \\ 1 & 0 \end{pmatrix} = \begin{pmatrix} 0 & 1 \\ 1 & 0 \end{pmatrix} = B$

Thus every even power of B is the identity matrix I and every odd power of B is the matrix B. Accordingly every power of B has zero entries, and so B is not regular.

(iii) C is not regular since it has a 1 on the main diagonal.

(iv) $$D^2 = \begin{pmatrix} 0 & 1 & 0 \\ \frac{1}{8} & \frac{5}{16} & \frac{9}{16} \\ \frac{1}{2} & \frac{1}{4} & \frac{1}{4} \end{pmatrix} \quad \text{and} \quad D^3 = \begin{pmatrix} \frac{1}{2} & \frac{1}{4} & \frac{1}{4} \\ \frac{5}{32} & \frac{41}{64} & \frac{13}{64} \\ \frac{1}{8} & \frac{5}{16} & \frac{9}{16} \end{pmatrix}$$

Since all the entries of D^3 are positive, D is regular.

MARKOV CHAINS

7.18. A student's study habits are as follows. If he studies one night, he is 70% sure not to study the next night. On the other hand, if he does not study one night, he is 60% sure not to study the next night as well. In the long run, how often does he study?

The states of the system are S (studying) and T (not studying). The transition matrix is

$$P = \begin{matrix} & S & T \\ S & .3 & .7 \\ T & .4 & .6 \end{matrix}$$

To discover what happens in the long run, we must find the unique fixed probability vector t of P. By Problem 7.13, $u = (.4, .7)$ is a fixed point of P and so $t = (\frac{4}{11}, \frac{7}{11})$ is the required probability vector. Thus in the long run the student studies $\frac{4}{11}$ of the time.

7.19. A psychologist makes the following assumptions concerning the behavior of mice subjected to a particular feeding schedule. For any particular trial 80% of the mice that went right on the previous experiment will go right on this trial, and 60% of those mice that went left on the previous experiment will go right on this trial. If 50% went right on the first trial, what would he predict for (i) the second trial, (ii) the third trial, (iii) the thousandth trial?

The states of the system are R (right) and L (left). The transition matrix is

$$P = \begin{matrix} & R & L \\ R & .8 & .2 \\ L & .6 & .4 \end{matrix}$$

The probability distribution for the first trial is $p = (.5, .5)$. To compute the probability distribution for the next step, i.e. the second trial, multiply p by the transition matrix P:

$$(.5, .5)\begin{pmatrix} .8 & .2 \\ .6 & .4 \end{pmatrix} = (.7, .3)$$

Thus on the second trial he predicts that 70% of the mice will go right and 30% will go left. To compute the probability distribution for the third trial, multiply that of the second trial by P:

$$(.7, .3)\begin{pmatrix} .8 & .2 \\ .6 & .4 \end{pmatrix} = (.74, .26)$$

Thus on the third trial he predicts that 74% of the mice will go right and 26% will go left.

We assume that the probability distribution for the thousandth trial is essentially the stationary probability distribution of the Markov chain, i.e. the unique fixed probability vector t of the transition matrix P. By Problem 7.13, $u = (.6, .2)$ is a fixed point of P and so $t = (\frac{3}{4}, \frac{1}{4}) = (.75, .25)$. Thus he predicts that, on the thousandth trial, 75% of the mice will go to the right and 25% will go to the left.

7.20. Given the transition matrix $P = \begin{pmatrix} 1 & 0 \\ \frac{1}{2} & \frac{1}{2} \end{pmatrix}$ with initial probability distribution $p^{(0)} = (\frac{1}{3}, \frac{2}{3})$. Define and find: (i) $p_{21}^{(3)}$, (ii) $p^{(3)}$, (iii) $p_2^{(3)}$.

(i) $p_{21}^{(3)}$ is the probability of moving from state a_2 to state a_1 in 3 steps. It can be obtained from the 3-step transition matrix P^3; hence first compute P^3:

$$P^2 = \begin{pmatrix} 1 & 0 \\ \frac{3}{4} & \frac{1}{4} \end{pmatrix}, \quad P^3 = \begin{pmatrix} 1 & 0 \\ \frac{7}{8} & \frac{1}{8} \end{pmatrix}$$

Then $p_{21}^{(3)}$ is the entry in the second row first column of P^3: $p_{21}^{(3)} = \frac{7}{8}$.

(ii) $p^{(3)}$ is the probability distribution of the system after three steps. It can be obtained by successively computing $p^{(1)}$, $p^{(2)}$ and then $p^{(3)}$:

$$p^{(1)} = p^{(0)}P = (\tfrac{1}{3}, \tfrac{2}{3}) \begin{pmatrix} 1 & 0 \\ \frac{1}{2} & \frac{1}{2} \end{pmatrix} = (\tfrac{2}{3}, \tfrac{1}{3})$$

$$p^{(2)} = p^{(1)}P = (\tfrac{2}{3}, \tfrac{1}{3}) \begin{pmatrix} 1 & 0 \\ \frac{1}{2} & \frac{1}{2} \end{pmatrix} = (\tfrac{5}{6}, \tfrac{1}{6})$$

$$p^{(3)} = p^{(2)}P = (\tfrac{5}{6}, \tfrac{1}{6}) \begin{pmatrix} 1 & 0 \\ \frac{1}{2} & \frac{1}{2} \end{pmatrix} = (\tfrac{11}{12}, \tfrac{1}{12})$$

However, since the 3-step transition matrix P^3 has already been computed in (i), $p^{(3)}$ can also be obtained as follows:

$$p^{(3)} = p^{(0)}P^3 = (\tfrac{1}{3}, \tfrac{2}{3}) \begin{pmatrix} 1 & 0 \\ \frac{7}{8} & \frac{1}{8} \end{pmatrix} = (\tfrac{11}{12}, \tfrac{1}{12})$$

(iii) $p_2^{(3)}$ is the probability that the process is in the state a_2 after 3 steps; it is the second component of the 3-step probability distribution $p^{(3)}$: $p_2^{(3)} = \frac{1}{12}$.

7.21. Given the transition matrix $P = \begin{pmatrix} 0 & \frac{1}{2} & \frac{1}{2} \\ \frac{1}{2} & \frac{1}{2} & 0 \\ 0 & 1 & 0 \end{pmatrix}$ and the initial probability distribution $p^{(0)} = (\frac{2}{3}, 0, \frac{1}{3})$. Find: (i) $p_{32}^{(2)}$ and $p_{13}^{(2)}$, (ii) $p^{(4)}$ and $p_3^{(4)}$, (iii) the vector that $p^{(0)}P^n$ approaches, (iv) the matrix that P^n approaches.

(i) First compute the 2-step transition matrix P^2:

$$P^2 = \begin{pmatrix} 0 & \frac{1}{2} & \frac{1}{2} \\ \frac{1}{2} & \frac{1}{2} & 0 \\ 0 & 1 & 0 \end{pmatrix} \begin{pmatrix} 0 & \frac{1}{2} & \frac{1}{2} \\ \frac{1}{2} & \frac{1}{2} & 0 \\ 0 & 1 & 0 \end{pmatrix} = \begin{pmatrix} \frac{1}{4} & \frac{3}{4} & 0 \\ \frac{1}{4} & \frac{1}{2} & \frac{1}{4} \\ \frac{1}{2} & \frac{1}{2} & 0 \end{pmatrix}$$

Then $p_{32}^{(2)} = \frac{1}{2}$ and $p_{13}^{(2)} = 0$, since these numbers refer to the entries in P^2.

(ii) To compute $p^{(4)}$, use the 2-step transition matrix P^2 and the initial probability distribution $p^{(0)}$:

$$p^{(2)} = p^{(0)}P^2 = (\tfrac{1}{3}, \tfrac{2}{3}, 0) \quad \text{and} \quad p^{(4)} = p^{(2)}P^2 = (\tfrac{1}{4}, \tfrac{7}{12}, \tfrac{1}{6})$$

Since $p_3^{(4)}$ is the third component of $p^{(4)}$, $p_3^{(4)} = \frac{1}{6}$.

(iii) By Theorem 7.3, $p^{(0)}P^n$ approaches the unique fixed probability vector t of P. To obtain t, first find any fixed vector $u = (x, y, z)$:

$$(x, y, z)\begin{pmatrix} 0 & \frac{1}{2} & \frac{1}{2} \\ \frac{1}{2} & \frac{1}{2} & 0 \\ 0 & 1 & 0 \end{pmatrix} = (x, y, z) \quad \text{or} \quad \begin{cases} \frac{1}{2}y = x \\ \frac{1}{2}x + \frac{1}{2}y + z = y \\ \frac{1}{2}x = z \end{cases}$$

Find any nonzero solution of the above system of equations. Set $z = 1$; then by the third equation $x = 2$, and by the first equation $y = 4$. Thus $u = (2, 4, 1)$ is a fixed point of P and so $t = (\frac{2}{7}, \frac{4}{7}, \frac{1}{7})$. In other words, $p^{(0)}P^n$ approaches $(\frac{2}{7}, \frac{4}{7}, \frac{1}{7})$.

(iv) P^n approaches the matrix T whose rows are each the fixed probability vector of P; hence

$$P^n \text{ approaches } \begin{pmatrix} \frac{2}{7} & \frac{4}{7} & \frac{1}{7} \\ \frac{2}{7} & \frac{4}{7} & \frac{1}{7} \\ \frac{2}{7} & \frac{4}{7} & \frac{1}{7} \end{pmatrix}.$$

7.22. A salesman's territory consists of three cities, A, B and C. He never sells in the same city on successive days. If he sells in city A, then the next day he sells in city B. However, if he sells in either B or C, then the next day he is twice as likely to sell in city A as in the other city. In the long run, how often does he sell in each of the cities?

The transition matrix of the problem is as follows:

$$P = \begin{matrix} & \begin{matrix} A & B & C \end{matrix} \\ \begin{matrix} A \\ B \\ C \end{matrix} & \begin{pmatrix} 0 & 1 & 0 \\ \frac{2}{3} & 0 & \frac{1}{3} \\ \frac{2}{3} & \frac{1}{3} & 0 \end{pmatrix} \end{matrix}$$

We seek the unique fixed probability vector t of the matrix P. First find any fixed vector $u = (x, y, z)$:

$$(x, y, z)\begin{pmatrix} 0 & 1 & 0 \\ \frac{2}{3} & 0 & \frac{1}{3} \\ \frac{2}{3} & \frac{1}{3} & 0 \end{pmatrix} = (x, y, z) \quad \text{or} \quad \begin{cases} \frac{2}{3}y + \frac{2}{3}z = x \\ x + \frac{1}{3}z = y \\ \frac{1}{3}y = z \end{cases}$$

Set, say, $z = 1$. Then by the third equation $y = 3$, and by the first equation $x = \frac{8}{3}$. Thus $u = (\frac{8}{3}, 3, 1)$. Also $3u = (8, 9, 3)$ is a fixed vector of P. Multiply $3u$ by $1/(8 + 9 + 3) = \frac{1}{20}$ to obtain the required fixed probability vector $t = (\frac{2}{5}, \frac{9}{20}, \frac{3}{20}) = (.40, .45, .15)$. Thus in the long run he sells 40% of the time in city A, 45% of the time in B and 15% of the time in C.

7.23. There are 2 white marbles in urn A and 3 red marbles in urn B. At each step of the process a marble is selected from each urn and the two marbles selected are interchanged. Let the state a_i of the system be the number i of red marbles in urn A. (i) Find the transition matrix P. (ii) What is the probability that there are 2 red marbles in urn A after 3 steps? (iii) In the long run, what is the probability that there are 2 red marbles in urn A?

(i) There are three states, a_0, a_1 and a_2 described by the following diagrams:

A	B	A	B	A	B
2 W	3 R	1 W, 1 R	1 W, 2 R	2 R	2 W, 1 R
a_0		a_1		a_2	

If the system is in state a_0, then a white marble must be selected from urn A and a red marble from urn B, so the system must move to state a_1. Accordingly, the first row of the transition matrix is $(0, 1, 0)$.

Suppose the system is in state a_1. It can move to state a_0 if and only if a red marble is selected from urn A and a white marble from urn B; the probability of that happening is $\frac{1}{2}\cdot\frac{1}{3} = \frac{1}{6}$. Thus $p_{10} = \frac{1}{6}$. The system can move from state a_1 to a_2 if and only if a white marble is selected from urn A and a red marble from urn B; the probability of that happening is $\frac{1}{2}\cdot\frac{2}{3} = \frac{1}{3}$. Thus $p_{12} = \frac{1}{3}$. Accordingly, the probability that the system remains in state a_1 is $p_{11} = 1 - \frac{1}{6} - \frac{1}{3} = \frac{1}{2}$. Thus the second row of the transition matrix is $(\frac{1}{6}, \frac{1}{2}, \frac{1}{3})$. (Note that p_{11} can also be obtained from the fact that the system remains in the state a_1 if either a white marble is drawn from each urn, probability $\frac{1}{2}\cdot\frac{1}{3} = \frac{1}{6}$, or a red marble is drawn from each urn, probability $\frac{1}{2}\cdot\frac{2}{3} = \frac{1}{3}$; thus $p_{11} = \frac{1}{6} + \frac{1}{3} = \frac{1}{2}$.)

Now suppose the system is in state a_2. A red marble must be drawn from urn A. If a red marble is selected from urn B, probability $\frac{1}{3}$, then the system remains in state a_2; and if a white marble is selected from urn B, probability $\frac{2}{3}$, then the system moves to state a_1. Note that the system can never move from state a_2 to the state a_0. Thus the third row of the transition matrix is $(0, \frac{2}{3}, \frac{1}{3})$. That is,

$$P = \begin{array}{c} \\ a_0 \\ a_1 \\ a_2 \end{array} \begin{array}{c} \begin{array}{ccc} a_0 & a_1 & a_2 \end{array} \\ \begin{pmatrix} 0 & 1 & 0 \\ \frac{1}{6} & \frac{1}{2} & \frac{1}{3} \\ 0 & \frac{2}{3} & \frac{1}{3} \end{pmatrix} \end{array}$$

(ii) The system began in state a_0, i.e. $p^{(0)} = (1, 0, 0)$. Thus:

$$p^{(1)} = p^{(0)}P = (0, 1, 0), \quad p^{(2)} = p^{(1)}P = (\tfrac{1}{6}, \tfrac{1}{2}, \tfrac{1}{3}), \quad p^{(3)} = p^{(2)}P = (\tfrac{1}{12}, \tfrac{23}{36}, \tfrac{5}{18})$$

Accordingly, the probability that there are 2 red marbles in urn A after 3 steps is $\frac{5}{18}$.

(iii) We seek the unique fixed probability vector t of the transition matrix P. First find any fixed vector $u = (x, y, z)$:

$$(x, y, z)\begin{pmatrix} 0 & 1 & 0 \\ \frac{1}{6} & \frac{1}{2} & \frac{1}{3} \\ 0 & \frac{2}{3} & \frac{1}{3} \end{pmatrix} = (x, y, z) \qquad \text{or} \qquad \begin{cases} \frac{1}{6}y = x \\ x + \frac{1}{2}y + \frac{2}{3}z = y \\ \frac{1}{3}y + \frac{1}{3}z = z \end{cases}$$

Set, say, $x = 1$. Then by the first equation $y = 6$, and by the third equation $z = 3$. Hence $u = (1, 6, 3)$. Multiply u by $1/(1+6+3) = \frac{1}{10}$ to obtain the required unique fixed probability vector $t = (.1, .6, .3)$. Thus, in the long run, 30% of the time there will be 2 red marbles in urn A.

Note that the long run probability distribution is the same as if the five marbles were placed in an urn and 2 were selected at random to put into urn A.

7.24. A player has \$2. He bets \$1 at a time and wins \$1 with probability $\frac{1}{2}$. He stops playing if he loses the \$2 or wins \$4. (i) What is the probability that he has lost his money at the end of, at most, 5 plays? (ii) What is the probability that the game lasts more than 7 plays?

This is random walk with absorbing barriers at 0 and 6 (see Examples 7.22 and 7.23). The transition matrix is

$$P = \begin{array}{c} \\ a_0 \\ a_1 \\ a_2 \\ a_3 \\ a_4 \\ a_5 \\ a_6 \end{array} \begin{array}{c} \begin{array}{ccccccc} a_0 & a_1 & a_2 & a_3 & a_4 & a_5 & a_6 \end{array} \\ \begin{pmatrix} 1 & 0 & 0 & 0 & 0 & 0 & 0 \\ \frac{1}{2} & 0 & \frac{1}{2} & 0 & 0 & 0 & 0 \\ 0 & \frac{1}{2} & 0 & \frac{1}{2} & 0 & 0 & 0 \\ 0 & 0 & \frac{1}{2} & 0 & \frac{1}{2} & 0 & 0 \\ 0 & 0 & 0 & \frac{1}{2} & 0 & \frac{1}{2} & 0 \\ 0 & 0 & 0 & 0 & \frac{1}{2} & 0 & \frac{1}{2} \\ 0 & 0 & 0 & 0 & 0 & 0 & 1 \end{pmatrix} \end{array}$$

with initial probability distribution $p^{(0)} = (0, 0, 1, 0, 0, 0, 0)$ since he began with \$2.

(i) We seek $p_0^{(5)}$, the probability that the system is in state a_0 after five steps. Compute the 5th step probability distribution $p^{(5)}$:

$$p^{(1)} = p^{(0)}P = (0, \tfrac{1}{2}, 0, \tfrac{1}{2}, 0, 0, 0) \qquad p^{(4)} = p^{(3)}P = (\tfrac{3}{8}, 0, \tfrac{5}{16}, 0, \tfrac{1}{4}, 0, \tfrac{1}{16})$$
$$p^{(2)} = p^{(1)}P = (\tfrac{1}{4}, 0, \tfrac{1}{2}, 0, \tfrac{1}{4}, 0, 0) \qquad p^{(5)} = p^{(4)}P = (\tfrac{3}{8}, \tfrac{5}{32}, 0, \tfrac{9}{32}, 0, \tfrac{1}{8}, \tfrac{1}{16})$$
$$p^{(3)} = p^{(2)}P = (\tfrac{1}{4}, \tfrac{1}{4}, 0, \tfrac{3}{8}, 0, \tfrac{1}{8}, 0)$$

Thus $p_0^{(5)}$, the probability that he has no money after 5 plays, is $\frac{3}{8}$.

(ii) Compute $p^{(7)}$: $p^{(6)} = p^{(5)}P = (\frac{29}{64}, 0, \frac{7}{32}, 0, \frac{13}{64}, 0, \frac{1}{8})$. $p^{(7)} = p^{(6)}P = (\frac{29}{64}, \frac{7}{64}, 0, \frac{27}{128}, 0, \frac{13}{128}, \frac{1}{8})$

The probability that the game lasts more than 7 plays, i.e. that the system is not in state a_0 or a_6 after 7 steps, is $\frac{7}{64} + \frac{27}{128} + \frac{13}{128} = \frac{27}{64}$.

7.25. Consider repeated tosses of a fair die. Let X_n be the maximum of the numbers occurring in the first n trials.

(i) Find the transition matrix P of the Markov chain. Is the matrix regular?

(ii) Find $p^{(1)}$, the probability distribution after the first toss.

(iii) Find $p^{(2)}$ and $p^{(3)}$.

(i) The state space of the Markov chain is $\{1, 2, 3, 4, 5, 6\}$. The transition matrix is

$$P = \begin{array}{c} \\ 1 \\ 2 \\ 3 \\ 4 \\ 5 \\ 6 \end{array} \begin{array}{c} \begin{array}{cccccc} 1 & 2 & 3 & 4 & 5 & 6 \end{array} \\ \begin{pmatrix} \frac{1}{6} & \frac{1}{6} & \frac{1}{6} & \frac{1}{6} & \frac{1}{6} & \frac{1}{6} \\ 0 & \frac{2}{6} & \frac{1}{6} & \frac{1}{6} & \frac{1}{6} & \frac{1}{6} \\ 0 & 0 & \frac{3}{6} & \frac{1}{6} & \frac{1}{6} & \frac{1}{6} \\ 0 & 0 & 0 & \frac{4}{6} & \frac{1}{6} & \frac{1}{6} \\ 0 & 0 & 0 & 0 & \frac{5}{6} & \frac{1}{6} \\ 0 & 0 & 0 & 0 & 0 & 1 \end{pmatrix} \end{array}$$

We obtain, for example, the third row of the matrix as follows. Suppose the system is in state 3, i.e. the maximum of the numbers occurring on the first n trials is 3. Then the system remains in state 3 if a 1, 2, or 3 occurs on the $(n+1)$-st trial; hence $p_{33} = \frac{3}{6}$. On the other hand, the system moves to state 4, 5 or 6, respectively, if a 4, 5 or 6 occurs on the $(n+1)$-st trial; hence $p_{34} = p_{35} = p_{36} = \frac{1}{6}$. The system can never move to state 1 or 2 since a 3 has occurred on one of the trials; hence $p_{31} = p_{32} = 0$. Thus the third row of the transition matrix is $(0, 0, \frac{3}{6}, \frac{1}{6}, \frac{1}{6}, \frac{1}{6})$. The other rows are obtained similarly.

The matrix is not regular since state 6 is absorbing, i.e. there is a 1 on the main diagonal in row 6.

(ii) On the first toss of the die, the state of the system X_1 is the number occurring; hence $p^{(1)} = (\frac{1}{6}, \frac{1}{6}, \frac{1}{6}, \frac{1}{6}, \frac{1}{6}, \frac{1}{6})$.

(iii) $p^{(2)} = p^{(1)}P = (\frac{1}{36}, \frac{3}{36}, \frac{5}{36}, \frac{7}{36}, \frac{9}{36}, \frac{11}{36})$. $p^{(3)} = p^{(2)}P = (\frac{1}{216}, \frac{7}{216}, \frac{19}{216}, \frac{37}{216}, \frac{61}{216}, \frac{91}{216})$.

7.26. Two boys b_1 and b_2 and two girls g_1 and g_2 are throwing a ball from one to the other. Each boy throws the ball to the other boy with probability $\frac{1}{2}$ and to each girl with probability $\frac{1}{4}$. On the other hand, each girl throws the ball to each boy with probability $\frac{1}{2}$ and never to the other girl. In the long run, how often does each receive the ball?

This is a Markov chain with state space $\{b_1, b_2, g_1, g_2\}$ and transition matrix

$$P = \begin{array}{c} \\ b_1 \\ b_2 \\ g_1 \\ g_2 \end{array} \begin{array}{c} \begin{array}{cccc} b_1 & b_2 & g_1 & g_2 \end{array} \\ \begin{pmatrix} 0 & \frac{1}{2} & \frac{1}{4} & \frac{1}{4} \\ \frac{1}{2} & 0 & \frac{1}{4} & \frac{1}{4} \\ \frac{1}{2} & \frac{1}{2} & 0 & 0 \\ \frac{1}{2} & \frac{1}{2} & 0 & 0 \end{pmatrix} \end{array}$$

We seek a fixed vector $u = (x, y, z, w)$ of P: $(x, y, z, w)P = (x, y, z, w)$. Set the corresponding components of uP equal to u to obtain the system

$$\begin{aligned} \tfrac{1}{2}y + \tfrac{1}{2}z + \tfrac{1}{2}w &= x \\ \tfrac{1}{2}x + \tfrac{1}{2}z + \tfrac{1}{2}w &= y \\ \tfrac{1}{4}x + \tfrac{1}{4}y &= z \\ \tfrac{1}{4}x + \tfrac{1}{4}y &= w \end{aligned}$$

We seek any nonzero solution. Set, say, $z = 1$; then $w = 1$, $x = 2$ and $y = 2$. Thus $u = (2, 2, 1, 1)$ and so the unique fixed probability of P is $t = (\frac{1}{3}, \frac{1}{3}, \frac{1}{6}, \frac{1}{6})$. Thus, in the long run, each boy receives the ball $\frac{1}{3}$ of the time and each girl $\frac{1}{6}$ of the time.

7.27. Prove Theorem 7.6: Let $P = (p_{ij})$ be the transition matrix of a Markov chain. If $p = (p_i)$ is the probability distribution of the system at some arbitrary time k, then pP is the probability distribution of the system one step later, i.e. at time $k+1$; hence pP^n is the probability distribution of the system n steps later, i.e. at time $k+n$. In particular, $p^{(1)} = p^{(0)}P$, $p^{(2)} = p^{(1)}P$, ... and also $p^{(n)} = p^{(0)}P^n$.

Suppose the state space is $\{a_1, a_2, \ldots, a_m\}$. The probability that the system is in state a_j at time k and then in state a_i at time $k+1$ is the product $p_j p_{ji}$. Thus the probability that the system is in state a_i at time $k+1$ is the sum

$$p_1 p_{1i} + p_2 p_{2i} + \cdots + p_m p_{mi} = \sum_{j=1}^{m} p_j p_{ji}$$

Thus the probability distribution at time $k+1$ is

$$p^* = \left(\sum_{j=1}^{m} p_j p_{j1}, \sum_{j=1}^{m} p_j p_{j2}, \ldots, \sum_{j=1}^{m} p_j p_{jm} \right)$$

However, this vector is precisely the product of the vector $p = (p_i)$ by the matrix $P = (p_{ij})$: $p^* = pP$.

7.28. Prove Theorem 7.5: Let P be the transition matrix of a Markov chain. Then the n-step transition matrix is equal to the nth power of P: $P^{(n)} = P^n$.

Suppose the system is in state a_i at, say, time k. We seek the probability $p_{ij}^{(n)}$ that the system is in state a_j at time $k+n$. Now the probability distribution of the system at time k, since the system is in state a_i, is the vector $e_i = (0, \ldots, 0, 1, 0, \ldots, 0)$ which has a 1 at the ith position and zeros everywhere else. By the preceding problem, the probability distribution at time $k+n$ is the product $e_i P^n$. But $e_i P^n$ is the ith row of the matrix P^n. Thus $P_{ij}^{(n)}$ is the jth component of the ith row of P^n, and so $P^{(n)} = P^n$.

MISCELLANEOUS PROBLEMS

7.29. The transition probabilities of a Markov chain can be represented by a diagram, called a *transition diagram*, where a positive probability p_{ij} is denoted by an arrow from the state a_i to the state a_j. Find the transition matrix of each of the following transition diagrams:

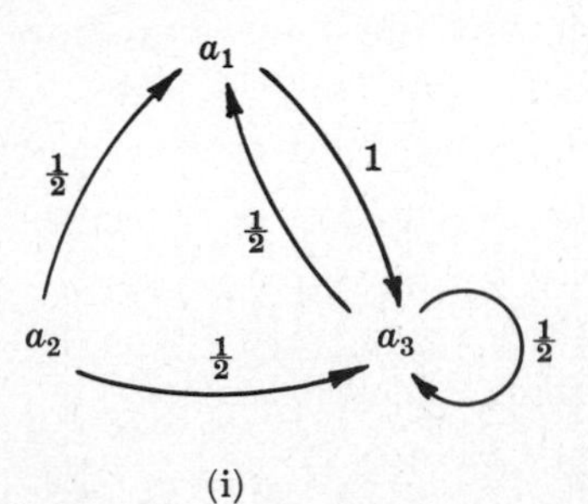

(i)

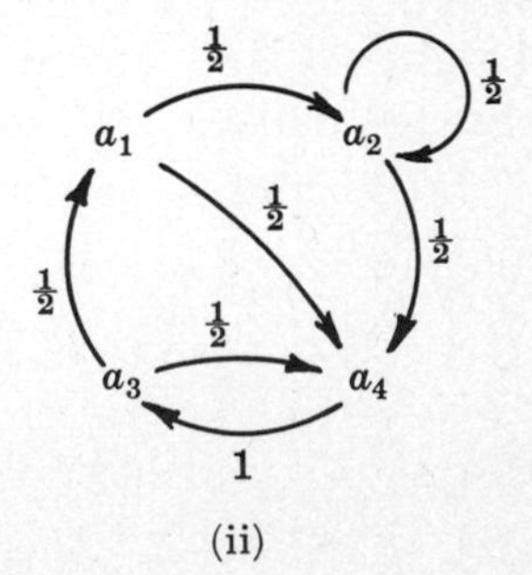

(ii)

(i) Note first that the state space is $\{a_1, a_2, a_3\}$ and so the transition matrix is of the form

$$P = \begin{array}{c} \\ a_1 \\ a_2 \\ a_3 \end{array} \begin{array}{c} a_1 \quad a_2 \quad a_3 \\ \left(\begin{array}{ccc} & & \\ & & \\ & & \end{array} \right) \end{array}$$

The ith row of the matrix is obtained by finding those arrows which emanate from a_i in the diagram; the number attached to the arrow from a_i to a_j is the jth component of the ith row. Thus the transition matrix is

$$P = \begin{array}{c} \\ a_1 \\ a_2 \\ a_3 \end{array} \begin{array}{c} \begin{array}{ccc} a_1 & a_2 & a_3 \end{array} \\ \begin{pmatrix} 0 & 0 & 1 \\ \frac{1}{2} & 0 & \frac{1}{2} \\ \frac{1}{2} & 0 & \frac{1}{2} \end{pmatrix} \end{array}$$

(ii) The state space is $\{a_1, a_2, a_3, a_4\}$. The transition matrix is

$$P = \begin{array}{c} \\ a_1 \\ a_2 \\ a_3 \\ a_4 \end{array} \begin{array}{c} \begin{array}{cccc} a_1 & a_2 & a_3 & a_4 \end{array} \\ \begin{pmatrix} 0 & \frac{1}{2} & 0 & \frac{1}{2} \\ 0 & \frac{1}{2} & 0 & \frac{1}{2} \\ \frac{1}{2} & 0 & 0 & \frac{1}{2} \\ 0 & 0 & 1 & 0 \end{pmatrix} \end{array}$$

7.30. Suppose the transition matrix of a Markov chain is as follows:

$$P = \begin{array}{c} \\ a_1 \\ a_2 \\ a_3 \\ a_4 \end{array} \begin{array}{c} \begin{array}{cccc} a_1 & a_2 & a_3 & a_4 \end{array} \\ \begin{pmatrix} \frac{1}{2} & \frac{1}{2} & 0 & 0 \\ \frac{1}{2} & \frac{1}{2} & 0 & 0 \\ \frac{1}{4} & \frac{1}{4} & \frac{1}{4} & \frac{1}{4} \\ \frac{1}{4} & \frac{1}{4} & \frac{1}{4} & \frac{1}{4} \end{pmatrix} \end{array}$$

Is the Markov chain regular?

Note that once the system enters the state a_1 or the state a_2, then it can never move to state a_3 or state a_4, i.e. the system remains in the state subspace $\{a_1, a_2\}$. Thus, in particular, $p_{13}^{(n)} = 0$ for every n and so every power P^n will contain a zero entry. Hence P is not regular.

7.31. Suppose m points on a circle are numbered respectively $1, 2, \ldots, m$ in a counterclockwise direction. A particle performs a "random walk" on the circle; it moves one step counterclockwise with probability p or one step clockwise with probability $q = 1 - p$. Find the transition matrix of this Markov chain.

The state space is $\{1, 2, \ldots, m\}$. The diagram to the right below can be used to obtain the transition matrix which appears to the left below.

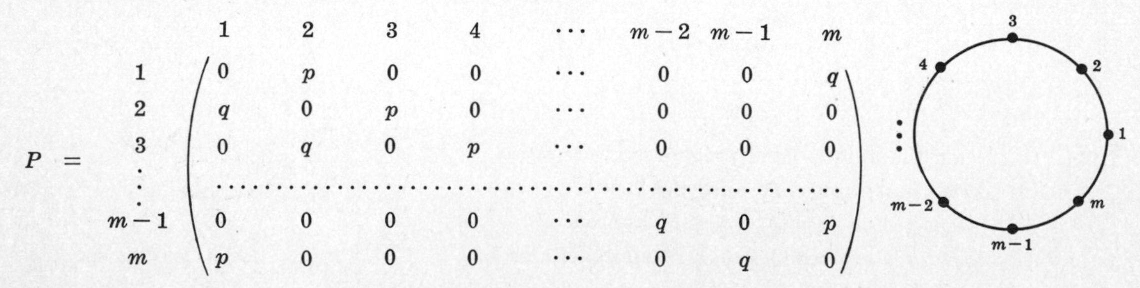

$$P = \begin{array}{c} \\ 1 \\ 2 \\ 3 \\ \vdots \\ m-1 \\ m \end{array} \begin{array}{c} \begin{array}{cccccccc} 1 & 2 & 3 & 4 & \cdots & m-2 & m-1 & m \end{array} \\ \begin{pmatrix} 0 & p & 0 & 0 & \cdots & 0 & 0 & q \\ q & 0 & p & 0 & \cdots & 0 & 0 & 0 \\ 0 & q & 0 & p & \cdots & 0 & 0 & 0 \\ \cdots & \cdots & \cdots & \cdots & \cdots & \cdots & \cdots & \cdots \\ 0 & 0 & 0 & 0 & \cdots & q & 0 & p \\ p & 0 & 0 & 0 & \cdots & 0 & q & 0 \end{pmatrix} \end{array}$$

Supplementary Problems

MATRIX MULTIPLICATION

7.32. Given $A = \begin{pmatrix} 1 & -2 & 3 \\ 4 & 1 & -1 \\ 5 & 2 & 3 \end{pmatrix}$. Find uA if (i) $u = (1, -3, 2)$, (ii) $u = (3, 0, -2)$, (iii) $u = (4, -1, -1)$.

7.33. Given $A = \begin{pmatrix} 1 & -1 & 4 \\ 3 & 1 & 5 \end{pmatrix}$ and $B = \begin{pmatrix} 2 & 1 \\ 6 & -3 \\ 1 & -2 \end{pmatrix}$. Find AB and BA.

7.34. Given $A = \begin{pmatrix} 2 & 2 \\ 3 & -1 \end{pmatrix}$. Find A^2 and A^3.

7.35. Given $A = \begin{pmatrix} 1 & 2 \\ 0 & 1 \end{pmatrix}$. Find A^n.

PROBABILITY VECTORS AND STOCHASTIC MATRICES

7.36. Which vectors are probability vectors?

(i) $(\frac{1}{4}, \frac{1}{2}, -\frac{1}{4}, \frac{1}{2})$ (ii) $(\frac{1}{2}, 0, \frac{1}{3}, \frac{1}{6}, \frac{1}{6})$ (iii) $(\frac{1}{12}, \frac{1}{2}, \frac{1}{6}, 0, \frac{1}{4})$.

7.37. Find a scalar multiple of each vector which is a probability vector:

(i) $(3, 0, 2, 5, 3)$ (ii) $(2, \frac{1}{2}, 0, \frac{1}{4}, \frac{3}{4}, 0, 1)$ (iii) $(\frac{1}{3}, 2, \frac{1}{2}, 0, \frac{1}{4}, \frac{2}{3})$.

7.38. Which matrices are stochastic?

(i) $\begin{pmatrix} 0 & 1 & 0 \\ \frac{1}{2} & \frac{1}{4} & \frac{1}{4} \end{pmatrix}$ (ii) $\begin{pmatrix} 1 & 0 \\ 0 & 1 \end{pmatrix}$ (iii) $\begin{pmatrix} 0 & 1 \\ \frac{1}{2} & \frac{1}{4} \end{pmatrix}$ (iv) $\begin{pmatrix} \frac{1}{2} & \frac{1}{2} \\ \frac{1}{2} & \frac{1}{2} \end{pmatrix}$ (v) $\begin{pmatrix} 0 & 1 \\ -\frac{1}{2} & \frac{3}{2} \end{pmatrix}$

REGULAR STOCHASTIC MATRICES AND FIXED PROBABILITY VECTORS

7.39. Find the unique fixed probability vector of each matrix:

(i) $\begin{pmatrix} \frac{2}{3} & \frac{1}{3} \\ \frac{2}{5} & \frac{3}{5} \end{pmatrix}$ (ii) $\begin{pmatrix} \frac{1}{4} & \frac{3}{4} \\ \frac{5}{6} & \frac{1}{6} \end{pmatrix}$ (iii) $\begin{pmatrix} .2 & .8 \\ .5 & .5 \end{pmatrix}$ (iv) $\begin{pmatrix} .7 & .3 \\ .6 & .4 \end{pmatrix}$

7.40. (i) Find the unique fixed probability vector t of $P = \begin{pmatrix} 0 & \frac{3}{4} & \frac{1}{4} \\ \frac{1}{2} & \frac{1}{2} & 0 \\ 0 & 1 & 0 \end{pmatrix}$.

(ii) What matrix does P^n approach? (iii) What vector does $(\frac{1}{4}, \frac{1}{4}, \frac{1}{2})P^n$ approach?

7.41. Find the unique fixed probability vector t of each matrix:

(i) $A = \begin{pmatrix} 0 & \frac{1}{2} & \frac{1}{2} \\ \frac{1}{3} & \frac{2}{3} & 0 \\ 0 & 1 & 0 \end{pmatrix}$ (ii) $B = \begin{pmatrix} 0 & 1 & 0 \\ \frac{1}{2} & 0 & \frac{1}{2} \\ \frac{1}{2} & \frac{1}{4} & \frac{1}{4} \end{pmatrix}$

7.42. (i) Find the unique fixed probability vector t of $P = \begin{pmatrix} 0 & \frac{1}{2} & \frac{1}{2} & 0 \\ \frac{1}{2} & \frac{1}{4} & 0 & \frac{1}{4} \\ 0 & 0 & 0 & 1 \\ 0 & \frac{1}{2} & 0 & \frac{1}{2} \end{pmatrix}$.

(ii) What matrix does P^n approach?

(iii) What vector does $(\frac{1}{4}, 0, \frac{1}{2}, \frac{1}{4})P^n$ approach?

(iv) What vector does $(\frac{1}{2}, 0, 0, \frac{1}{2})P^n$ approach?

7.43. (i) Given that $t = (\frac{1}{2}, 0, \frac{1}{4}, \frac{1}{4})$ is a fixed point of a stochastic matrix P, is P regular?
(ii) Given that $t = (\frac{1}{4}, \frac{1}{4}, \frac{1}{4}, \frac{1}{4})$ is a fixed point of a stochastic matrix P, is P regular?

7.44. Which of the stochastic matrices are regular?

$$\text{(i)} \begin{pmatrix} \frac{1}{2} & \frac{1}{4} & \frac{1}{4} \\ 0 & 1 & 0 \\ \frac{1}{2} & 0 & \frac{1}{2} \end{pmatrix} \quad \text{(ii)} \begin{pmatrix} \frac{1}{2} & \frac{1}{2} & 0 \\ \frac{1}{2} & \frac{1}{2} & 0 \\ \frac{1}{4} & \frac{1}{4} & \frac{1}{2} \end{pmatrix} \quad \text{(iii)} \begin{pmatrix} 0 & 0 & 1 \\ \frac{1}{2} & 0 & \frac{1}{2} \\ 0 & 1 & 0 \end{pmatrix}$$

7.45. Show that $(cf + ce + de,\ af + bf + ae,\ ad + bd + bc)$ is a fixed point of the matrix

$$P = \begin{pmatrix} 1-a-b & a & b \\ c & 1-c-d & d \\ e & f & 1-e-f \end{pmatrix}$$

MARKOV CHAINS

7.46. A man's smoking habits are as follows. If he smokes filter cigarettes one week, he switches to nonfilter cigarettes the next week with probability .2. On the other hand, if he smokes nonfilter cigarettes one week, there is a probability of .7 that he will smoke nonfilter cigarettes the next week as well. In the long run, how often does he smoke filter cigarettes?

7.47. A gambler's luck follows a pattern. If he wins a game, the probability of winning the next game is .6. However, if he loses a game, the probability of losing the next game is .7. There is an even chance that the gambler wins the first game.
(i) What is the probability that he wins the second game?
(ii) What is the probability that he wins the third game?
(iii) In the long run, how often will he win?

7.48. For a Markov chain, the transition matrix is $P = \begin{pmatrix} \frac{1}{2} & \frac{1}{2} \\ \frac{3}{4} & \frac{1}{4} \end{pmatrix}$ with initial probability distribution $p^{(0)} = (\frac{1}{4}, \frac{3}{4})$. Find: (i) $p_{21}^{(2)}$; (ii) $p_{12}^{(2)}$; (iii) $p^{(2)}$; (iv) $p_1^{(2)}$; (v) the vector $p^{(0)}P^n$ approaches; (vi) the matrix P^n approaches.

7.49. For a Markov chain, the transition matrix is $P = \begin{pmatrix} \frac{1}{2} & 0 & \frac{1}{2} \\ 1 & 0 & 0 \\ \frac{1}{4} & \frac{1}{2} & \frac{1}{4} \end{pmatrix}$ and the initial probability distribution is $p^{(0)} = (\frac{1}{2}, \frac{1}{2}, 0)$. Find (i) $p_{13}^{(2)}$, (ii) $p_{23}^{(2)}$, (iii) $p^{(2)}$, (iv) $p_1^{(2)}$.

7.50. Each year a man trades his car for a new car. If he has a Buick, he trades it for a Plymouth. If he has a Plymouth, he trades it for a Ford. However, if he has a Ford, he is just as likely to trade it for a new Ford as to trade it for a Buick or a Plymouth. In 1955 he bought his first car which was a Ford.
(i) Find the probability that he has a (*a*) 1957 Ford, (*b*) 1957 Buick, (*c*) 1958 Plymouth, (*d*) 1958 Ford.
(ii) In the long run, how often will he have a Ford?

7.51. There are 2 white marbles in urn A and 4 red marbles in urn B. At each step of the process a marble is selected from each urn, and the two marbles selected are interchanged. Let X_n be the number of red marbles in urn A after n interchanges. (i) Find the transition matrix P. (ii) What is the probability that there are 2 red marbles in urn A after 3 steps? (iii) In the long run, what is the probability that there are 2 red marbles in urn A?

7.52. Solve the preceding problem in the case that there are 3 white marbles in urn A and 3 red marbles in urn B.

7.53. A fair coin is tossed until 3 heads occur in a row. Let X_n be the length of the sequence of heads ending at the nth trial. (See Example 7.24.) What is the probability that there are at least 8 tosses of the coin?

7.54. A player has 3 dollars. At each play of a game, he loses one dollar with probability $\frac{3}{4}$ but wins two dollars with probability $\frac{1}{4}$. He stops playing if he has lost his 3 dollars or he has won at least 3 dollars.

(i) Find the transition matrix of the Markov chain.

(ii) What is the probability that there are at least 4 plays to the game?

7.55. The diagram on the right shows four compartments with doors leading from one to another. A mouse in any compartment is equally likely to pass through each of the doors of the compartment. Find the transition matrix of the Markov chain.

MISCELLANEOUS PROBLEMS

7.56. Find the transition matrix corresponding to each transition diagram:

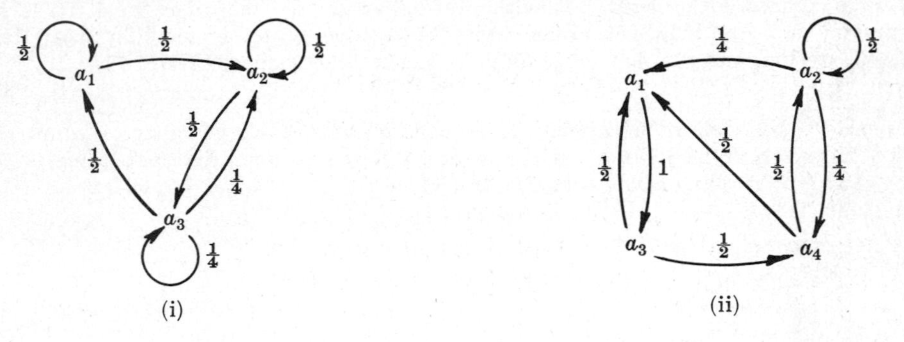

7.57. Draw a transition diagram for each transition matrix:

(i) $P = \begin{matrix} & a_1 & a_2 \\ a_1 & \frac{1}{2} & \frac{1}{2} \\ a_2 & \frac{1}{3} & \frac{2}{3} \end{matrix}$ (ii) $P = \begin{matrix} & a_1 & a_2 & a_3 \\ a_1 & 0 & \frac{1}{2} & \frac{1}{2} \\ a_2 & \frac{1}{4} & \frac{1}{4} & \frac{1}{2} \\ a_3 & 0 & \frac{1}{2} & \frac{1}{2} \end{matrix}$

7.58. Consider the vector $e_i = (0, \ldots, 0, 1, 0, \ldots, 0)$ which has a 1 at the ith position and zeros elsewhere. Show that e_iA is the ith row of the matrix A (whenever the product is defined).

Answers to Supplementary Problems

7.32. (i) $(-1, -1, 12)$, (ii) $(-7, -10, 3)$, (iii) $(-5, -11, 10)$

7.33. $AB = \begin{pmatrix} 0 & -4 \\ 17 & -10 \end{pmatrix}$, $BA = \begin{pmatrix} 5 & -1 & 13 \\ -3 & -9 & 9 \\ -5 & -3 & -6 \end{pmatrix}$

7.34. $A^2 = \begin{pmatrix} 10 & 2 \\ 3 & 7 \end{pmatrix}$, $A^3 = \begin{pmatrix} 26 & 18 \\ 27 & -1 \end{pmatrix}$

7.35. $A^n = \begin{pmatrix} 1 & 2n \\ 0 & 1 \end{pmatrix}$

7.36. Only (iii).

7.37. (i) (3/13, 0, 2/13, 5/13, 3/13)
(ii) (8/18, 2/18, 0, 1/18, 3/18, 0, 4/18)
(iii) (4/45, 24/45, 6/45, 0, 3/45, 8/45)

7.38. Only (ii) and (iv).

7.39. (i) (6/11, 5/11), (ii) (10/19, 9/19), (iii) (5/13, 8/13), (iv) $(\frac{2}{3}, \frac{1}{3})$

7.40. (i) $t = (4/13, 8/13, 1/13)$, (iii) $t = (4/13, 8/13, 1/13)$

7.41. (i) $t = (2/9, 6/9, 1/9)$, (ii) $t = (5/15, 6/15, 4/15)$

7.42. (i) $t = (2/11, 4/11, 1/11, 4/11)$, (iii) t, (iv) t

7.43. (i) No, (ii) not necessarily, e.g. $P = \begin{pmatrix} 1 & 0 & 0 & 0 \\ 0 & 1 & 0 & 0 \\ 0 & 0 & 1 & 0 \\ 0 & 0 & 0 & 1 \end{pmatrix}$

7.44. Only (iii)

7.46. 60% of the time

7.47. (i) 9/20, (ii) 87/200, (iii) 3/7 of the time

7.48. (i) 9/16, (ii) 3/8, (iii) (37/64, 27/64), (iv) 37/64, (v) (.6, .4), (vi) $\begin{pmatrix} .6 & .4 \\ .6 & .4 \end{pmatrix}$

7.49. (i) 3/8, (ii) 1/2, (iii) (7/16, 2/16, 7/16), (iv) 7/16

7.50. (i) (*a*) 4/9, (*b*) 1/9, (*c*) 7/27, (*d*) 16/27. (ii) 50% of the time

7.51. (i) $P = \begin{pmatrix} 0 & 1 & 0 \\ \frac{1}{8} & \frac{1}{2} & \frac{3}{8} \\ 0 & \frac{1}{2} & \frac{1}{2} \end{pmatrix}$ (ii) 3/8 (iii) 2/5

7.52. (i) $P = \begin{pmatrix} 0 & 1 & 0 & 0 \\ \frac{1}{9} & \frac{4}{9} & \frac{4}{9} & 0 \\ 0 & \frac{4}{9} & \frac{4}{9} & \frac{1}{9} \\ 0 & 0 & 1 & 0 \end{pmatrix}$ (ii) 32/81 (iii) 9/20

7.53. 81/128

7.54. (i) $P = \begin{pmatrix} 1 & 0 & 0 & 0 & 0 & 0 & 0 \\ \frac{3}{4} & 0 & 0 & \frac{1}{4} & 0 & 0 & 0 \\ 0 & \frac{3}{4} & 0 & 0 & \frac{1}{4} & 0 & 0 \\ 0 & 0 & \frac{3}{4} & 0 & 0 & \frac{1}{4} & 0 \\ 0 & 0 & 0 & \frac{3}{4} & 0 & 0 & \frac{1}{4} \\ 0 & 0 & 0 & 0 & \frac{3}{4} & 0 & \frac{1}{4} \\ 0 & 0 & 0 & 0 & 0 & 0 & 1 \end{pmatrix}$ (ii) 27/64

7.55. $P = \begin{pmatrix} 0 & \frac{2}{3} & 0 & \frac{1}{3} \\ \frac{2}{3} & 0 & \frac{1}{3} & 0 \\ 0 & \frac{1}{2} & 0 & \frac{1}{2} \\ \frac{1}{2} & 0 & \frac{1}{2} & 0 \end{pmatrix}$

7.56. (i) $\begin{pmatrix} \frac{1}{2} & \frac{1}{2} & 0 \\ 0 & \frac{1}{2} & \frac{1}{2} \\ \frac{1}{2} & \frac{1}{4} & \frac{1}{4} \end{pmatrix}$ (ii) $\begin{pmatrix} 0 & 0 & 1 & 0 \\ \frac{1}{4} & \frac{1}{2} & 0 & \frac{1}{4} \\ \frac{1}{2} & 0 & 0 & \frac{1}{2} \\ \frac{1}{2} & \frac{1}{2} & 0 & 0 \end{pmatrix}$

7.57. (i)

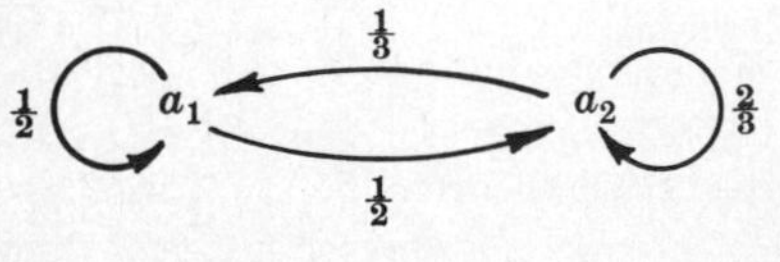

$\frac{1}{2}$ a_1 $\frac{1}{3}$ $\frac{1}{2}$ a_2 $\frac{2}{3}$

(ii)

a_1 $\frac{1}{4}$ $\frac{1}{2}$ $\frac{1}{2}$ $\frac{1}{2}$ a_2 a_3 $\frac{1}{4}$ $\frac{1}{2}$ $\frac{1}{2}$

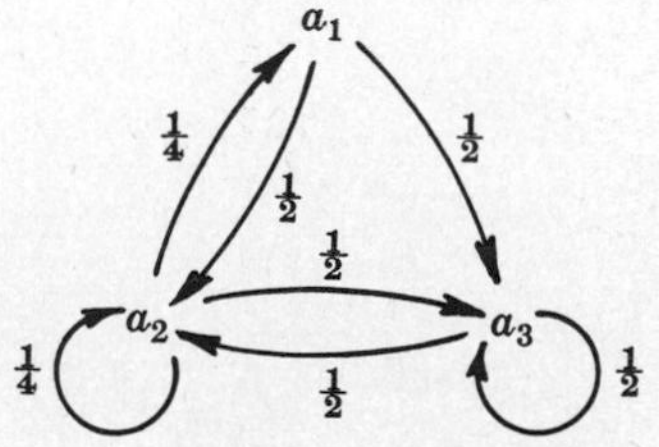

INDEX